Sebastian Andres

Diffusion Processes with Reflection

Sebastian Andres

Diffusion Processes with Reflection

Südwestdeutscher Verlag für Hochschulschriften

Impressum/Imprint (nur für Deutschland/ only for Germany)
Bibliografische Information der Deutschen Nationalbibliothek: Die Deutsche Nationalbibliothek verzeichnet diese Publikation in der Deutschen Nationalbibliografie; detaillierte bibliografische Daten sind im Internet über http://dnb.d-nb.de abrufbar.
Alle in diesem Buch genannten Marken und Produktnamen unterliegen warenzeichen-, marken- oder patentrechtlichem Schutz bzw. sind Warenzeichen oder eingetragene Warenzeichen der jeweiligen Inhaber. Die Wiedergabe von Marken, Produktnamen, Gebrauchsnamen, Handelsnamen, Warenbezeichnungen u.s.w. in diesem Werk berechtigt auch ohne besondere Kennzeichnung nicht zu der Annahme, dass solche Namen im Sinne der Warenzeichen- und Markenschutzgesetzgebung als frei zu betrachten wären und daher von jedermann benutzt werden dürften.

Verlag: Südwestdeutscher Verlag für Hochschulschriften Aktiengesellschaft & Co. KG
Dudweiler Landstr. 99, 66123 Saarbrücken, Deutschland
Telefon +49 681 37 20 271-1, Telefax +49 681 37 20 271-0, Email: info@svh-verlag.de
Zugl.: Berlin, TU, Diss., 2009

Herstellung in Deutschland:
Schaltungsdienst Lange o.H.G., Zehrensdorfer Str. 11, D-12277 Berlin
Books on Demand GmbH, Gutenbergring 53, D-22848 Norderstedt
Reha GmbH, Dudweiler Landstr. 99, D- 66123 Saarbrücken
ISBN: 978-3-8381-0928-2

Imprint (only for USA, GB)
Bibliographic information published by the Deutsche Nationalbibliothek: The Deutsche Nationalbibliothek lists this publication in the Deutsche Nationalbibliografie; detailed bibliographic data are available in the Internet at http://dnb.d-nb.de.
Any brand names and product names mentioned in this book are subject to trademark, brand or patent protection and are trademarks or registered trademarks of their respective holders. The use of brand names, product names, common names, trade names, product descriptions etc. even without
a particular marking in this works is in no way to be construed to mean that such names may be regarded as unrestricted in respect of trademark and brand protection legislation and could thus be used by anyone.

Publisher:
Südwestdeutscher Verlag für Hochschulschriften Aktiengesellschaft & Co. KG
Dudweiler Landstr. 99, 66123 Saarbrücken, Germany
Phone +49 681 37 20 271-1, Fax +49 681 37 20 271-0, Email: info@svh-verlag.de

Copyright © 2008 Südwestdeutscher Verlag für Hochschulschriften Aktiengesellschaft & Co. KG and licensors
All rights reserved. Saarbrücken 2008

Produced in USA and UK by:
Lightning Source Inc., 1246 Heil Quaker Blvd., La Vergne, TN 37086, USA
Lightning Source UK Ltd., Chapter House, Pitfield, Kiln Farm, Milton Keynes, MK11 3LW, GB
BookSurge, 7290 B. Investment Drive, North Charleston, SC 29418, USA
ISBN: 978-3-8381-0928-2

Preface

Es gibt nichts Praktischeres als eine gute Theorie.
IMMANUEL KANT

In modern probability theory diffusion processes with reflection arise in various manners. As an example let us consider the reflected Brownian motion, that is a Brownian motion on the positive real line with reflection in zero. There are several possibilities to construct this process, the simplest and most intuitive one is certainly the following: Given a standard Brownian motion B, starting in some $x \geq 0$, a reflected Brownian motion is obtained by taking the absolute value $|B|$. By the Itô-Tanaka formula we have

$$|B_t| = x + \int_0^t \operatorname{sign}(B_s)\, dB_s + L_t,$$

where $(L_t)_{t \geq 0}$ is the local time of B in zero, i.e. it is continuous, nondecreasing and $\operatorname{supp}(dL) \subseteq \{t \geq 0 : B_t = 0\}$. Note that $\beta_t := \int_0^t \operatorname{sign}(X_s)\, dB_s$ is again a Brownian motion by Lévy's characterization theorem.

Another possibility to construct a reflected Brownian motion is to apply the Skorohod reflection principle. Given a standard Brownian motion W there exists a unique pair (X, L) such that

$$X_t = x + W_t + L_t, \qquad x \geq 0,\, t \geq 0, \tag{0.1}$$

where $X_t \geq 0$ for all t, L starts in zero and is continuous, monotone nondecreasing and $\operatorname{supp}(dL) \subseteq \{t \geq 0 : X_t = 0\}$, i.e. it increases only at those times, when X is zero. Moreover, the local time L is given by $L_t = [-x - \inf_{s \leq t} W_s]^+$. Here the reflected Brownian motion (X_t) is constructed by adding the local time L. Equation (0.1) is a simple example for a so called Skorohod-SDE. Since β defined above is a Brownian motion, one can easily check that

$$X \stackrel{d}{=} |B|.$$

A much more abstract and analytic method to introduce the reflected Brownian motion is to define it via its Dirichlet form. The reflected Brownian motion is the diffusion associated with the Dirichlet form

$$\mathcal{E}(f, g) = \frac{1}{2} \int_0^\infty f'(x)\, g'(x)\, dx, \qquad f, g \in W^{1,2}([0, \infty)).$$

i

Another meaningful class of stochastic processes involving some kind of reflection are Bessel processes. The Bessel process ρ^δ with Bessel dimension $\delta > 0$ is again a diffusion process taking nonnegative values, which is associated with the Dirichlet form

$$\mathcal{E}^\delta(f,g) = C_\delta \int_0^\infty f'(x)\,g'(x)\,x^{\delta-1}\,dx, \qquad f,g \in W^{1,2}([0,\infty), x^{\delta-1}dx),$$

C_δ denoting a positive constant. Usually Bessel processes are introduced by defining the square of a Bessel process (see e.g. Chapter XI in [56]). For integer dimensions, i.e. for $\delta \in [1,\infty) \cap \mathbb{N}$, ρ^δ can be obtained by taking the Euclidian norm of a δ-dimensional Brownian motion. In particular, the one-dimensional Bessel process ρ^1 is just the reflected Brownian motion discussed above. For $\delta > 1$, the Bessel process ρ^δ starting in some $x \geq 0$ can be described as the unique solution of the SDE

$$\rho_t^\delta = x + \frac{\delta-1}{2}\int_0^t \frac{1}{\rho_s^\delta}\,ds + B_t. \qquad (0.2)$$

The drift term appearing in this equation effects a repulsion, which is strong enough to force the process to stay nonnegative, so that no further reflection term is needed. Finally, for $0 < \delta < 1$ the situation is much more complicated, because the drift term is now attracting and not repulsive. As a consequence, a representation of ρ^δ in terms of an SDE like (0.1) or (0.2) is not possible in this case, because ρ^δ is known not to be a semi-martingale if $\delta < 1$ (see e.g. Section 6 in [57]). Nevertheless, ρ^δ admits a Fukushima decomposition

$$\rho_t^\delta = B_t + (\delta - 1)H_t, \qquad t \geq 0,$$

as the sum of a Brownian motion and a zero energy process $(\delta - 1)H$, which also produces the reflection in this case, see [13] for details.

In this thesis, diffusion processes will appear in nearly every fashion mentioned so far. The thesis consists of two parts. The first part contains some pathwise differentiability results for Skorohod SDEs. In the second part a particle approximation of the Wasserstein diffusion is established, where the approximating process can be interpeted as a system of interacting Bessel process with small Bessel dimension. The second part is a joint work with Max von Renesse, TU Berlin.

Pathwise Differentiability for SDEs with Reflection

Consider some closed bounded connected domain G in $\mathbb{R}^d$, $d \geq 2$, and for any starting point $x \in G$ the following SDE of the Skorohod type:

$$X_t(x) = x + \int_0^t b(X_r(x))\,dr + w_t + \int_0^t \gamma(X_r(x))\,dl_r(x), \qquad t \geq 0,$$

$$X_t(x) \in G, \quad dl_t(x) \geq 0, \quad \int_0^\infty \mathbb{1}_{G\setminus \partial G}(X_t(x))\,dl_t(x) = 0, \qquad t \geq 0,$$

where w is a d-dimensional Brownian motion and b is a continuously differentiable drift. Furthermore, for every $x \in \partial G$, $\gamma(x)$ denotes the direction of reflection at x, for instance in the case of normal reflection $\gamma(x)$ is the inner normal field on ∂G. Finally, $l(x)$ the local time of $X(x)$ in ∂G, i.e. it increases only at those times, when $X(x)$ is at ∂G.

In the first part of this thesis we investigate the question whether the mapping $x \mapsto X_t(x)$, $t > 0$, is differentiable almost surely and whether one can characterize the derivatives in order to derive a Bismut formula. Deuschel and Zambotti have solved this problem in [26] for the domain $G = [0, \infty)^d$ and several related questions have already been considered in [6].

In Chapter 1 (cf. [8]), we consider the case where G is a convex polyhedron and where the directions of reflection along each face are constant but possibly oblique. The differentiability is obtained for every time t up to the first time when the process X hits two of the faces simultaneously. Similarly to the results in [26], the derivatives evolve according to a linear ordinary differential equation, when the process X is in the interior of the domain, and they are projected to the tangent space, when it hits the boundary. In particular, when X reaches the end of an excursion interval, the derivative process jumps in the direction of the corresponding direction of reflection. This evolution becomes rather non-trival due to the complicated structure of the set of times, when the process is at the boundary, which is known to a.s. a closed set of zero Lebesgue measure without isolated points.

Chapter 2 (cf. [7]) deals with the case where G is a bounded smooth domain with normal reflection at the boundary. The additional difficulties appearing here are caused on one hand by the presence of curvature and on the other hand by the fact that one cannot apply anymore the technical lemma, on which the proofs in [26] and [8] are based, dealing with the time a Brownian path attains its minimum. As a result we obtain an analogous time evolution for the derivatives as described above. In contrast to the results in [26], this evolution gives not a complete characterization of the derivatives in this case. Therefore, a further SDE-like equation is established, characterizing the derivatives in the coordinates w.r.t. a moving frame.

Particle Approximation of the Wasserstein Diffusion

The Wasserstein diffusion, recently constructed in [66] by using abstract Dirichlet form methods, is roughly speaking a reversible Markov process, taking values in the set $\mathcal{P}$ of probability measures on the unit interval, whose intrinsic metric is given by the L^2-Wasserstein distance. In order to improve the intuitive and mathematical understanding of that process, we establish a reversible particle system consisting of N diffusing particles on the unit interval, whose associated empirical measure process converges weakly to the Wasserstein diffusion in the high-density limit, while we have to assume Markov uniqueness for the Dirichlet form, which induces the Wasserstein diffusion (cf. [9]).

This approximating particle system can be interpreted as a system of coupled, two-sided Bessel processes with small Bessel dimension. In particular, for large N the drift term, which appears in the SDE describing formally the particle system, is attracting and not repulsive. In analogy to the Bessel process it is not contained in the class of Euclidian semi-martingales.

A detailed analysis of the approximating system, in particular Feller properties, will be subject of Chapter 3 (cf. [10]) based on harmonic analysis on weighted Sobolev spaces. A crucial step here is to verify the Muckenhoupt condition for the reversible measure, which will imply the doubling property and a uniform local Poincaré inequality. From this one can derive the Feller property using an abstract result by Sturm [59].

To obtain the convergence result one has to prove tightness and in a second step one has to identify the limit. While tightness easily follows by some well-established tightness-criteria, the identification of the limit is more difficult. The proof uses the parametrization of $\mathcal{P}$ in terms of right-continuous quantile functions. Then, the convergence of the empirical measure process is equivalent to the convergence of the process of the corresponding quantile functions. This convergence is then obtained by proving that the associated Dirichlet forms converge in the Mosco sense to the limiting Dirichlet form, where we use the generalized version of Mosco convergence with varying base spaces established by Kuwai and Shioya (cf. [50]). The verification of the Mosco conditions is based on an integration by parts formula for the invariant measure of the limiting process (Prop. 7.3 in [66]) and on the fact that the logarithmic derivatives of the equilibrium distributions converge in an appropiate L^2-sense. Moreover, the assumption of Markov uniqueness is a crucial ingredient in the proof of the Mosco convergence.

Acknowledgement

Foremost, I sincerely thank my advisor Prof. Jean-Dominique Deuschel for giving me the opportunity to write a PhD thesis in an exciting area of probability theory and for his continual support during the last years.

I am also very indebted to Max von Renesse, my co-author of the second part, for his patience and for innumerable instructive discussions. Moreover, I would like to thank Lorenzo Zambotti for accepting the task of being the co-examiner of this thesis, for his hospitality during my stay in Paris as well as for many fruitful discussions and helpful hints and comments.

Many thanks to Erwin Bolthausen and the research groups at the University of Zurich and ETH Zurich for a very pleasant stay in spring 2008. I am grateful to a lot of further mathematicians for useful comments, in particular Krzysztof Burdzy, Michael Röckner, Alain-Sol Sznitman and many others.

During my time as PhD student, I very much appreciated the inspiring atmosphere in the stochastic community at the Technical University, Humboldt University and the Weierstrass Institute. I thank my friends and colleages and all members of the group for the great time I spent here. Finally, I am very grateful to my family and friends for their steady support throughout the years.

Financial support by the Deutsche Forschungsgemeinschaft through the International Research Training Group "Stochastic Models of Complex Processes" is gratefully acknowledged.

Contents

I   Pathwise Differentiability for SDEs with Reflection   1

1   SDEs in a convex Polyhedron with oblique Reflection   3
  1.1   Introduction   3
  1.2   Model and Main Result   4
  1.3   Martingale Problem and Neumann Condition   8
  1.4   Proof of the Main Result 12
        1.4.1   Continuity w.r.t. the Initial Condition 12
        1.4.2   Computation of the Local Times 13
        1.4.3   Computation of the Difference Quotients 15
        1.4.4   Proof of the Differentiability 18

2   SDEs in a Smooth Domain with normal Reflection   21
  2.1   Introduction 21
  2.2   Main Results and Preliminaries 22
        2.2.1   General Notation 22
        2.2.2   Skorohod SDE 23
        2.2.3   Main Results 24
        2.2.4   Localization 26
        2.2.5   Example: Processes in the Unit Ball 28
  2.3   Proof of the Main Result 28
        2.3.1   Lipschitz Continuity w.r.t. the Initial Datum 28
        2.3.2   Convergence of the Local Time Processes 32
        2.3.3   Proof of the Differentiability 36
        2.3.4   The Neumann Condition 46

II   Particle Approximation of the Wasserstein Diffusion   49

3   The Approximating Particle System   51
  3.1   Introduction and Main Results 51
  3.2   Dirichlet Form and Integration by Parts Formula 54
  3.3   Feller Property 56
        3.3.1   Feller Properties of Y 57

3.3.2 Feller Properties of Y inside a Box 60
3.3.3 Feller Properties of X^N 65
3.4 Semi-Martingale Properties 69

4 Weak Convergence to the Wasserstein Diffusion   **73**
4.1 Introduction 73
4.2 Main Result 76
4.3 Tightness 76
4.4 Identification of the Limit 78
4.4.1 The $\mathcal{G}$-Parameterization 78
4.4.2 Finite Dimensional Approximation of Dirichlet Forms in Mosco Sense . . . 80
4.4.3 Condition Mosco II' 82
4.4.4 Condition Mosco I 86
4.4.5 Proof of Proposition 4.4 93
4.5 A non-convex $(1+1)$-dimensional $\nabla\phi$-interface model 95

A Conditional Expectation of the Dirichlet Process   **97**

Bibliography   **101**

Part I

Pathwise Differentiability for SDEs with Reflection

Chapter 1

SDEs in a convex Polyhedron with oblique Reflection

1.1 Introduction

We consider a Markov process with continuous sample paths, characterized as the strong solution of a stochastic differential equation (SDE) of the Skorohod type, where the domain G is a convex polyhedron in $\mathbb{R}^d$, i.e. G is the intersection of a finite number of half spaces. The process is driven by a d-dimensional standard Brownian motion and a drift term, whose coefficient function is supposed to be continuously differentiable and Lipschitz continuous. At the boundary of the polyhedron it reflects instantaneously, the possibly oblique direction of reflection being constant along each face.

Let $G = \bigcap_{i=1}^{N} G_i$, where each G_i is a closed half space with inward normal n_i. The direction of reflection on the faces ∂G_i will be denoted by constant vectors v_i. As an example one might think of the process of a Brownian motion in an infinite two-dimensional wedge, established by Varadhan and Williams in [64] (see Figure 1.1). Existence and uniqueness for solutions of SDEs with oblique reflecting boundary conditions on polyhedral domains are ensured by a result of Dupuis and Ishii in [29]. The study of such SDEs is motivated by several applications: For instance, these processes arise as diffusion approximations of storage systems or of single-server queues in heavy traffic (see e.g. Section 8.4 in [20] for details).

In this chapter we show that the solution of the Skorohod SDE is pathwise differentiable with respect to the deterministic initial value and to characterize the pathwise derivatives up to time τ, when at least two faces of G are hit simultaneously for the first time. This is an addition to the results of [26], where Deuschel and Zambotti considered such a differentiability problem for SDEs on the domain $G = [0, \infty)^d$ with normal reflection at the boundary. Our proceeding will be quite similar to that in [26], in particular we shall use the same technical lemma dealing with the minimum of a Brownian path (see Lemma 1 in [26]). The resulting derivatives are described in terms of an ODE-like equation. When the process is away from the boundary, they evolve according to a simple linear ordinary differential equation, and when it hits the boundary, they have a discontinuity; more precisely, they are projected to the tangent space and jump in direction of the corresponding reflection vector (cf. Section 1.3 below). In addition, we provide

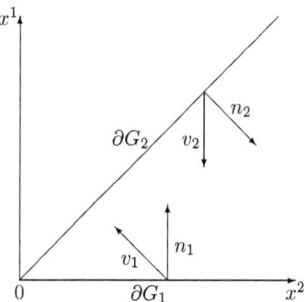

Figure 1.1: Two-dimensional Wegde with oblique Reflection

a Bismut-Elworthy formula for the gradient of the transition semigroup of the process which is stopped in τ (see Corollary 1.5 below).

A crucial step in the proof of the differentiability result is to show that the solution of the Skorohod SDE depends Lipschitz continuously on the initial value. To do this we shall apply a criterion given in [28]. In particular, we have to ensure that a certain static geometric property holds (cf. Assumption 2.1 in [28]), so that an additional restriction to the directions of reflection is needed.

Our result is similar to a system, which has been introduced by Airault in [1] in order to develop probabilistic representations for the solutions of linear PDE systems with mixed Dirichlet-Neumann conditions on a regular domain in $\mathbb{R}^n$. However, in contrast to [1] we study pathwise differentiability properties of a process with reflection following [26], but with possibly oblique reflection.

The material presented in this chapter is contained in [8]. The chapter is organized as follows: In Section 1.2 we state the main result and in Section 1.4 we prove it. In Section 1.3 we investigate the results in detail, while we establish a martingale problem connected with the derivatives and we check the Neumann condition.

1.2 Model and Main Result

In this chapter we denote by $\|.\|$ the Euclidian norm, by $\langle .,.\rangle$ the canonical scalar product and by $e = (e^1, \ldots, e^d)$ the standard basis in $\mathbb{R}^d$, $d \geq 2$. We consider processes on the domain G, which is a convex polyhedron, i.e. $G \subseteq \mathbb{R}^d$ takes the form $G = \bigcap_{i=1}^{N} G_i$, where each $G_i := \{x : \langle x, n_i \rangle \geq c_i\}$ is a closed half space with inward normal n_i and intercept c_i. The boundary of the polyhedron consists of the sides $\partial G_i = \{x : \langle x, n_i \rangle = c_i\}$ and with each side ∂G_i we associate a constant, possibly oblique direction of reflection v_i, pointing into the interior of the polyhedron. We always adopt the convention that the directions v_i are normalized such that $\langle v_i, n_i \rangle = 1$. For every

1.2 MODEL AND MAIN RESULT

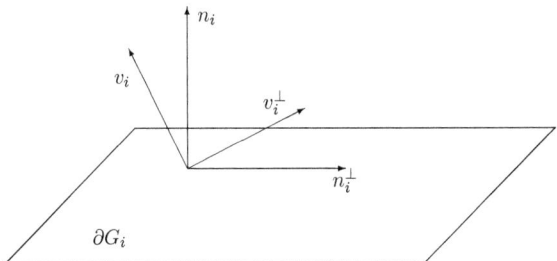

Figure 1.2: Choice of $n_i^\perp$ and $v_i^\perp$

$i \in \{1, \ldots, N\}$, let $v_i^\perp, n_i^\perp \in \text{span}\{n_i, v_i\}$ be such that

$$\langle v_i, v_i^\perp \rangle = \langle n_i, n_i^\perp \rangle = 0, \qquad \langle n_i^\perp, v_i^\perp \rangle = \langle n_i, v_i \rangle = 1, \qquad \langle n_i, v_i^\perp \rangle > 0, \qquad (1.1)$$

which implies $\langle v_i, n_i^\perp \rangle = -\langle v_i^\perp, n_i \rangle$ (cf. Figure 1.2). Furthermore, for every i let $n_i^1 := n_i$, $n_i^2 = n_i^\perp$ and $(n_i^k)_{k=3,\ldots,d}$ be such that $\{n_i^1, \ldots, n_i^d\}$ is an orthonormal basis of $\mathbb{R}^d$.

To ensure Lipschitz continuity (see Lemma 1.11 below) and pathwise existence and uniqueness, a further assumption on the directions of reflection is needed, namely that either

$$n_i = v_i \qquad \text{or} \qquad a_i \langle n_i, v_i \rangle > \sum_{j \neq i} a_j |\langle n_i, v_j \rangle| \qquad (1.2)$$

for some positive constants a_i and for all i (cf. Theorem 2.1 in [28]).

The set of continuous real-valued functions on G is denoted by $C(G)$, and $C_b(G)$ denotes the set of those functions in $C(G)$ that are bounded on G. For each $k \in \mathbb{N}$, $C^k(G)$ denotes the set of real-valued functions that are k-times continuously differentiable in some domain containing G, and $C_b^k(G)$ denotes the set of those functions in $C^k(G)$ that are bounded and have bounded partial derivatives up to order k. Furthermore, for $f \in C^1(G)$ we denote by ∇f the gradient of f and in the case where f is $\mathbb{R}^d$-valued by Df the Jacobi matrix. Finally, Δ denotes the Laplace differential operator on $C^2(G)$ and $D_v := \langle v, \nabla \rangle$ the directional derivative operator associated with the direction $v \in G$.

For any starting point $x \in G$, we consider the following stochastic differential equation of the Skorohod type:

$$\begin{aligned} X_t(x) &= x + \int_0^t b(X_r(x))\, dr + w_t + \sum_i v_i\, l_t^i(x), \qquad t \geq 0, \\ X_t(x) &\in G, \quad dl_t^i(x) \geq 0, \quad \int_0^\infty \mathbb{1}_{G \setminus \partial G_i}(X_t(x))\, dl_t^i(x) = 0, \qquad t \geq 0,\ i \in \{1, \ldots, N\}, \end{aligned} \qquad (1.3)$$

where w is a d-dimensional Brownian motion on a complete probability space $(\Omega, \mathcal{F}, \mathbb{P})$. For every i, $l^i(x)$ denotes the local time of $X(x)$ in ∂G_i, i.e. it increases only at those times, when $X(x)$ is at the boundary ∂G_i. The components $b^i : G \to \mathbb{R}$ of b are supposed to be in $C^1(G)$ and Lipschitz continuous. Then, existence and uniqueness of strong solutions of (1.3) are guaranteed in the case of normal reflection by the results of [60], since G is convex, and in the case of oblique reflection by [29], since by condition (1.2) the assumptions of Case 2 in [29] are fulfilled (cf. Remark 3.1 in [29]).

Notice that there is one degree of freedom in defining the local times in the Skorohod SDE (1.3): Setting $\tilde{l}^i(x) = h_i l^i(x)$ for any real constants $h_i > 0$, $\tilde{l}^i(x)$ satisfies the conditions in (1.3) as well. Thus, it is possible to replace $v_i l^i(x)$ by $h_i^{-1} v_i \tilde{l}^i(x)$ in the Skorohod SDE. Consequently, the norm of the reflection vectors v_i does not affect the Skorohod equation, so that the vectors can be thought to be normalized. However, we shall use the normalization $\langle v_i, n_i \rangle = 1$ chosen above, to simplify the computations in the sequel.

Furthermore, by the Girsanov Theorem there exists a probability measure $\tilde{\mathbb{P}}(x)$, which is equivalent to $\mathbb{P}$, such that the process

$$W_t^i(x) := \int_0^t b^i(X_r(x)) \, dr + w_t^i, \qquad t \geq 0, \, i \in \{1, \ldots, d\}, \tag{1.4}$$

is a d-dimensional Brownian motion under $\tilde{\mathbb{P}}(x)$. Next we define the stopping time τ by

$$\tau := \inf\{t \geq 0 : X_t(x) \in \partial G_i \cap \partial G_j, \, i \neq j\}, \qquad x \in G, \tag{1.5}$$

to be the first time, when the process hits at least two of the faces simultaneously. The following simple example shows that even under the assumption in (1.2) τ can a.s. be infinite and finite as well.

Example 1.1. Let $G = \mathbb{R}_+^2$, i.e. G is a two-dimensional wedge with angle $\frac{\pi}{2}$ and inward normals $n_i = e_i$, $i \in \{1, 2\}$ (cf. Figure 1.1). We choose $v_1 = n_1$ and $v_2 = (-\tan \theta, 1)$, where $\theta \in (-\frac{\pi}{4}, \frac{\pi}{4})$ denotes the angle between n_2 and v_2, such that the vector v_2 points towards the corner if θ is positive. Then, the assumption in (1.2) holds with $a_1 = a_2 = 1$. From Theorem 2.2 in [64] we know that

$$\mathbb{P}[\tau < \infty] = \begin{cases} 0 & \text{if } \theta \leq 0, \\ 1 & \text{if } \theta > 0, \end{cases}$$

for any starting point $x \in G\backslash\{0\}$. Nevertheless, τ has infinite expectation for every $\theta \in (-\frac{\pi}{2}, \frac{\pi}{2})$ (see Corollary 2.3 in [64]).

We set

$$C^i := \{s \geq 0 : X_s(x) \in \partial G_i\}, \qquad r_i(t) := \sup(C^i \cap [0, t]), \quad i \in \{1, \ldots, N\},$$

with the convention $\sup \emptyset := 0$, and furthermore $C := \bigcup_{i=1}^N C^i$ and $r(t) := \max_{\{i=1,\ldots,N\}} r_i(t)$. Then, for every i, $C^i \cap [0, \tau)$ is known to be a.s. a closed set of zero Lebesgue measure without

1.2 Model and Main Result

isolated points (closed relative to $[0, \tau)$) and $t \mapsto r_i(t)$ is locally constant and right-continuous. For $t \in [0, \tau)$ we define

$$s(t) := \begin{cases} 0 & \text{if } t < \inf C, \\ i & \text{if } r(t) = r_i(t), \end{cases}$$

i.e. $s(t) = i$ if the last hit of the boundary before time t was in ∂G_i, and $s(t) = 0$ if up to time t the process has not hit the boundary yet. Let $(A_n)_n$ be the family of connected components of $[0, \tau) \backslash C$. A_n is open, so that there exists $q_n \in A_n \cap \mathbb{Q}$, $n \in \mathbb{N}$. Let $a_n := \inf A_n$ be the starting points and $b_n := \sup A_n$ be the endpoints of the excursion intervals. Finally, let $\tau_0 := 0$ and $\tau_0 < \tau_1 < \tau_2 < \ldots < \tau$ be the jump times of $t \mapsto s(t)$ on $[0, \tau)$, i.e. at every time τ_ℓ, $\ell > 0$, the process X has crossed the polyhedron from one face to another one. The following theorem gives a representation of the derivatives of X in terms of an ODE-like equation:

Theorem 1.2. *For all $t \in [0, \tau)$ and $x \in G$ a.s. the mapping $y \mapsto X_t(y)$ is differentiable at x and, setting $\eta_t := D_v X_t(x) = \lim_{\varepsilon \to 0} (X_t(x + \varepsilon v) - X_t(x))/\varepsilon$, $v \in \mathbb{R}^d$, there exists a right-continuous modification of η such that a.s.*

$$\begin{aligned} \eta_t &= v + \int_0^t Db(X_r(x)) \cdot \eta_r \, dr, & \text{if } s(t) = 0, \\ \eta_t &= \langle \eta_{r(t)-}, v_i^\perp \rangle n_i^\perp + \sum_{k=3}^d \langle \eta_{r(t)-}, n_i^k \rangle n_i^k + \int_{r(t)}^t Db(X_r(x)) \cdot \eta_r \, dr, & \text{if } s(t) = i. \end{aligned} \quad (1.6)$$

The proof of Theorem 1.2 is postponed to Section 1.4. If we consider the case $G = \mathbb{R}_+^d$ and normal reflection at the boundary, i.e. $v_i = n_i = e_i$, the result corresponds to that of Theorem 1 in [26]. In the special case where $N = d$ and the normals n_i form an orthonormal basis of $\mathbb{R}^d$, it is also possible to provide a random walk representation for the derivatives, which is very similar to that in [26], by using essentially the same arguments as in the proof of Theorem 1 and Proposition 1 in [26].

Remark 1.3. If $x \in \partial G_i$ for any i, $t = 0$ is a.s. an accumulation point of C and we have $r(t) > 0$ a.s. for every $t > 0$. Therefore, in that case $\eta_0 = v$ and $\eta_{0+} = \langle v, v_i^\perp \rangle n_i^\perp + \sum_{k=3}^d \langle v, n_i^k \rangle n_i^k$, i.e. there is discontinuity at $t = 0$.

Remark 1.4. The equation (1.6) does not characterize the derivatives, since it does not admit a unique solution. Indeed, if the process (η_t) solves (1.6), then the process $(1 + l_t(x))\eta_t$, $t \geq 0$, also does. A characterizing equation for the derivatives is given Theorem 1.6 below.

As soon as pathwise differentiability is established, we can immediately provide a Bismut-Elworthy formula: Define $X_t^\tau(x) := X_t(x) \mathbb{1}_{\{t < \tau\}}$ and for all $f \in C_b(G)$ the associated transition semigroup $P_t f(x) := \mathbb{E}[f(X_t^\tau(x))]$, $x \in G$, $t > 0$. Setting $\eta_t^{ij} := \partial X_t^i(x)/\partial x^j$ for $t \in [0, \tau)$ and $\eta^{ij} := 0$ on $[\tau, \infty)$, $i, j \in \{1, \ldots, d\}$, we get

Corollary 1.5. *For all $f \in C_b(G)$, $t > 0$ and $x \in G$:*

$$\frac{\partial}{\partial x^i} P_t f(x) = \frac{1}{t} \mathbb{E}\left[f(X_t^\tau(x)) \int_0^t \sum_{k=1}^d \eta_r^{ki} \, dw_r^k \right], \qquad i \in \{1, \ldots, d\}, \quad (1.7)$$

and if $f \in C_b^1(G)$:

$$\frac{\partial}{\partial x^i} P_t f(x) = \sum_{k=1}^{d} \mathbb{E}\left[\frac{\partial f}{\partial x^k}(X_t^\tau(x))\, \eta_t^{ki}\right], \qquad i \in \{1,\ldots,d\}. \qquad (1.8)$$

Proof. Formula (1.8) is straightforward from the differentiability statement in Theorem 1.2 and the chain rule. For formula (1.7) see the proof of Theorem 2 in [26]. □

Finally, we give another equation, which characterizes η locally. For that purpose, we set

$$y_t^k := \begin{cases} \langle \eta_t, e^k \rangle & \text{if } s(t) = 0, \\ \langle \eta_t, n_i^k \rangle & \text{if } s(t) = i, \end{cases} \quad \text{and} \quad c_t(k,l) := \begin{cases} \langle e^k, Db(X_t(x)) \cdot e^l \rangle & \text{if } s(t) = 0, \\ \langle n_i^k, Db(X_t(x)) \cdot n_i^l \rangle & \text{if } s(t) = i. \end{cases}$$

Theorem 1.6. *There exists a right-continuous modification of η and y, respectively, such that y is characterized as the unique solution of*

$$y_t^k = \langle v, e^k \rangle + \int_0^t \sum_{l=1}^{d} c_r(k,l)\, y_r^l\, dr, \qquad k \in \{1,\ldots,d\}, \qquad \text{if } t < \inf C \qquad (1.9a)$$

and

$$y_t^1 = \int_{r_i(t)}^{t} \sum_{l=1}^{d} c_r(1,l)\, y_r^l\, dr,$$

$$y_t^2 = \langle v_i^\perp, n_i \rangle\, y_{\tau_\ell-}^1 + y_{\tau_\ell-}^2 + \int_{\tau_\ell}^{t} \sum_{l=1}^{d} c_r(2,l)\, y_r^l\, dr + \langle v_i^\perp, n_i \rangle \int_{\tau_\ell}^{r_i(t)} \sum_{l=1}^{d} c_r(2,l)\, y_r^l\, dr, \qquad (1.9b)$$

$$y_t^k = y_{\tau_\ell-}^k + \int_{\tau_\ell}^{t} \sum_{l=1}^{d} c_r(k,l)\, y_r^l\, dr, \qquad k \in \{3,\ldots,d\},$$

if $t \geq \inf C$, ℓ such that $t \in [\tau_\ell, \tau_{\ell+1})$ and with $s(t) = i$. If $x \in \partial G_i$ for some i, i.e. $\inf C = 0$, we also need to specify $y_{0-}^k = y_{\tau_0-}^k := \langle v, n_i^k \rangle$, $k = 1,\ldots,d$.

Again the proof is postponed to Section 1.4.

1.3 Martingale Problem and Neumann Condition

In this section we investigate the derivatives of X, established in Theorem 1.2, in detail. Let v be arbitrary but fixed. From the representation of the derivatives in (1.6) it is obvious that $(\eta_t)_{0 \leq t < \tau}$ evolves according to a linear differential equation, when the process X is in the interior of the polyhedron, and that it is projected to the tangent space, when X is at the boundary. Furthermore, if X hits the boundary ∂G_i at some time t_i and we have $r(t_i-) \neq r(t_i)$, i.e. t_i is the endpoint b_n of an excursion interval A_n for some $n \in \mathbb{N}$, then also η has a discontinuity at t_i and jumps as follows:

$$\eta_{t_i} = \langle \eta_{t_i-}, v_i^\perp \rangle n_i^\perp + \sum_{k=3}^{d} \langle \eta_{t_i-}, n_i^k \rangle n_i^k.$$

1.3 MARTINGALE PROBLEM AND NEUMANN CONDITION

Recall that $\{n_i^k;\, k=1,\dots,d\}$ is an orthonormal basis of $\mathbb{R}^d$, so that $\eta_{t_i-} = \sum_{k=1}^d \langle \eta_{t_i-}, n_i^k\rangle n_i^k$ and

$$\eta_{t_i} - \eta_{t_i-} = \langle \eta_{t_i-}, v_i^\perp - n_i^\perp\rangle n_i^\perp - \langle \eta_{t_i-}, n_i\rangle n_i = -\langle \eta_{t_i-}, n_i\rangle v_i, \tag{1.10}$$

where the last equality follows from Lemma 1.7 below. Consequently, we observe that at each time, when X reaches the boundary ∂G_i, η is projected to the tangent space, since $\langle \eta_{t_i}, n_i\rangle = 0$, and jumps in direction of v_i or $-v_i$, respectively. Finally, if $X_{t_i}(x) \in \partial G_i$ and $t \mapsto r(t)$ is continuous in $t = t_i$, there is also a projection of η, but since in this case η_{t_i-} is in the tangent space, the projection has no effect and η is continuous at time t_i.

Lemma 1.7. *For all $i \in \{1,\dots,N\}$ and $\eta \in \mathbb{R}^d$:*

$$\langle \eta, v_i^\perp - n_i^\perp\rangle n_i^\perp - \langle \eta, n_i\rangle n_i = -\langle \eta, n_i\rangle v_i.$$

Proof. By the choice of $v_i^\perp$ and $n_i^\perp$ in (1.1) we have $v_i = n_i + \langle v_i, n_i^\perp\rangle n_i^\perp$ and $v_i^\perp = \langle v_i^\perp, n_i\rangle n_i + n_i^\perp$, which is equivalent to

$$\langle v_i, n_i^\perp\rangle n_i^\perp = v_i - n_i, \qquad v_i^\perp - n_i^\perp = -\langle v_i, n_i^\perp\rangle n_i.$$

Hence,

$$\langle \eta, v_i^\perp - n_i^\perp\rangle n_i^\perp - \langle \eta, n_i\rangle n_i = -\langle \eta, n_i\rangle \langle v_i, n_i^\perp\rangle n_i^\perp - \langle \eta, n_i\rangle n_i = -\langle \eta, n_i\rangle(v_i - n_i) - \langle \eta, n_i\rangle n_i$$
$$= -\langle \eta, n_i\rangle v_i.$$

$\square$

From the observations above it becomes clear that the process $(X_t(x), \eta_t)_{0 \le t < \tau}$ is Markovian with state space $G \times \mathbb{R}^d$. Next we want to provide the infinitesimal generator for this Markov process. For that purpose we define the operator $\mathcal{L}$ as follows: Let the domain $\mathcal{D}(\mathcal{L})$ be that set of continuous bounded functions F on $G \times \mathbb{R}^d$ satisfying the following conditions:

i) For every $\eta \in \mathbb{R}^d$, $F(\cdot, \eta) \in C_b^2(G)$ and the Neumann boundary condition holds:

$$D_{v_i} F(\cdot, \eta)(x) = 0 \qquad \text{for } x \in \partial G_i, \quad i \in \{1,\dots,N\}. \tag{1.11}$$

ii) For every $x \in G$, we have $F(x, \cdot) \in C_b^1(\mathbb{R}^d)$, i.e. bounded and continuously differentiable with bounded partial derivatives, satisfying the following boundary conditions: If $x \in \partial G_i$ for all $\eta \in \mathbb{R}^d$:

$$F(x, \eta) = F(x, \eta - \langle \eta, n_i\rangle v_i), \qquad D_{v_i} F(x, \cdot)(\eta) = 0. \tag{1.12}$$

Note that by the jump behaviour of η, provided in (1.10), and by the boundary condition (1.12) we have for every $F \in \mathcal{D}(\mathcal{L})$ and $t < \tau$:

$$F(X_t, \eta_t) = F(X_t, \eta_{t-}). \tag{1.13}$$

Finally, the operator $\mathcal{L}$ is defined by:
$$\mathcal{L}F(x,\eta) := \mathcal{L}^1 F(.,\eta)(x) + \mathcal{L}^2 F(x,.)(\eta), \quad F \in \mathcal{D}(\mathcal{L}),$$
where
$$\mathcal{L}^1 F(.,\eta)(x) := \frac{1}{2}\Delta F(.,\eta)(x) + \sum_{i=1}^{d} b^i(x)\frac{\partial F}{\partial x^i}(.,\eta)(x),$$
$$\mathcal{L}^2 F(x,.)(\eta) := \sum_{i=1}^{d}\left(\sum_{k=1}^{d}\frac{\partial b^i}{\partial x^k}(x)\,\eta^k\right)\frac{\partial F}{\partial \eta^i}(x,.)(\eta).$$

Proposition 1.8. *For $F \in \mathcal{D}(\mathcal{L})$,*
$$F(X_t(x),\eta_t) - F(x,\eta_0) - \int_0^t \mathcal{L}F(X_s,\eta_s)\,ds, \quad t < \tau,$$
is a martingale.

Proof. Since $(X_t(x),\eta_t)_{t<\tau}$ is a semimartingale (see Proposition 1.9 below), we may apply Itô's formula for right-continuous semimartingales (see e.g. Section II.7 in [55]) to obtain

$$F(X_t(x),\eta_t) - F(x,\eta_0)$$
$$= \int_0^t \nabla_x F(X_s(x),\eta_s)\,dX_s(x) + \int_0^t \nabla_\eta F(X_s(x),\eta_s)\,d\eta_s + \frac{1}{2}\int_0^t \Delta F(.,\eta_s)(X_s(x))\,ds$$
$$+ \sum_{0<s\leq t}\{F(X_s(x),\eta_s) - F(X_s(x),\eta_{s-}) - \nabla_\eta F(X_s(x),\eta_{s-}) \cdot (\eta_s - \eta_{s-})\}$$
$$= m_t + \int_0^t \mathcal{L}F(X_s(x),\eta_s)\,ds + \sum_{i=1}^{N}\int_0^t D_{v_i}F(X_s(x),\eta_s)\,dl_s^i(x)$$
$$+ \sum_{0<s\leq t} F(X_s(x),\eta_s) - F(X_s(x),\eta_{s-}) - \sum_{0<s\leq t}\nabla_\eta F(X_s(x),\eta_{s-}) \cdot (\eta_s - \eta_{s-}),$$

where (m_t) is a martingale. Clearly, the third and the fourth term vanish by the boundary conditions (1.11) and (1.13). Using (1.10) the last term can be rewritten as

$$-\sum_{0<s\leq t}\sum_{m=1}^{N}\nabla_\eta F(X_s(x),\eta_{s-}) \cdot (\eta_s - \eta_{s-})\,\mathbb{1}_{\{X_s(x)\in\partial G_m\}}$$
$$= \sum_{0<s\leq t}\sum_{m=1}^{N}\langle \eta_{s-},n_m\rangle \sum_{i=1}^{d}D_{v_m}F(X_s(x),.)(\eta_{s-})\,\mathbb{1}_{\{X_s(x)\in\partial G_m\}},$$

which is equal to zero by (1.12). $\square$

Since $(X_t(x),\eta_t)_{t<\tau}$ is Markovian, we can conclude from Proposition 1.8 that the restriction of its generator to $\mathcal{D}(\mathcal{L})$ coincides with $\mathcal{L}$. Note that Proposition 1.8 does possibly not give a complete description of the law of the process (X,η) because it only states existence and not uniqueness for solutions of the associated martingale problem.

1.3 Martingale Problem and Neumann Condition

Proposition 1.9. $(\eta_t)_{0 \le t < \tau}$ *is a process of bounded variation.*

Proof. It suffices to show that the sizes of the jumps of η on $[0, t]$ are summable for every $t < \tau$. On one hand, η has a jump at time τ_ℓ for each ℓ. Since the number of crossing through the polyhedron from one face to another one up to time t is finite a.s., it is enough to show that

$$\sum_{r \in (\tau_\ell, \tau_{\ell+1} \wedge t)} \|\eta_r - \eta_{r-}\| < \infty, \quad \text{for every } \tau_\ell \le t.$$

Let $\tau_\ell \le t$ and i be such that $s(r) = i$ for all $r \in (\tau_\ell, \tau_{\ell+1})$. Setting $\mathcal{A}_\ell := \{n \in \mathbb{N} : A_n \subset (\tau_\ell, \tau_{\ell+1} \wedge t)\}$, we use (1.10) and (1.6) to obtain

$$\sum_{r \in (\tau_\ell, \tau_{\ell+1})} \|\eta_r - \eta_{r-}\| = \sum_{n \in \mathcal{A}_\ell} \|\eta_{b_n} - \eta_{b_n-}\| = \sum_{n \in \mathcal{A}_\ell} \|\langle \eta_{b_n-}, n_i \rangle v_i \|$$

$$\le \|v_i\| \sum_{n \in \mathcal{A}_\ell} \int_{a_n}^{b_n} |\langle n_i, Db(X_r(x)) \cdot \eta_r \rangle| \, dr$$

$$\le c \sum_{n \in \mathcal{A}_\ell} (b_n - a_n) \le c(t - \tau_\ell),$$

for some positive constant c. The uniform boundedness of η_r in r, which is used here, will follow from Lemma 1.11 below. $\square$

Finally, the following induction argument gives another confirmation of our results, namely they imply that the Neumann condition holds for X.

Corollary 1.10. *Let again* $X_t^\tau(x) := X_t(x) \mathbb{1}_{\{t < \tau\}}$. *Then, for all* $f \in C_b(G)$ *and* $t > 0$, *the transition semigroup* $P_t f(x) := \mathbb{E}[f(X_t^\tau(x))]$, $x \in G$, *satisfies the Neumann condition at* ∂G:

$$x \in \partial G_i \implies D_{v_i} P_t f(x) = 0.$$

Proof. Let $x \in \partial G_i$. By a density argument it is sufficient to consider bounded functions f, which are continuously differentiable and have bounded derivatives. Then, for each $t > 0$ we obtain by dominated convergence and the chain rule:

$$D_{v_i} P_t f(x) = \mathbb{E}\left[\nabla f(X_t(x)) D_{v_i} X_t(x) \mathbb{1}_{\{t \in [0, \tau)\}}\right].$$

Thus, it suffices to show

$$D_{v_i} X_t(x) = 0, \quad \forall 0 < t < \tau.$$

For this purpose it is enough to show that $y_t = 0$ for all $t \in [\tau_\ell, \tau_{\ell+1})$ for every ℓ with $v = v_i$. We shall use induction on ℓ. Recall that $\inf C = 0$. Since $v = v_i$, using Theorem 1.6 we obtain that

$$y_0^2 = \langle v_i^\perp, n_i \rangle \langle v_i, n_i \rangle + \langle v_i, n_i^\perp \rangle = 0 \quad \text{and} \quad y_0^k = \langle v_i, n_i^k \rangle = 0, \quad k = 3, \ldots, d.$$

Thus, for every $t \in [0, \tau_1)$,

$$y_t^1 = \int_{r_i(t)}^{t} \sum_{l=1}^{d} c_r(1,l)\, y_r^l\, dr,$$

$$y_t^2 = \int_{0}^{t} \sum_{l=1}^{d} c_r(2,l)\, y_r^l\, dr + \langle v_i^\perp, n_i \rangle \int_{0}^{r_i(t)} \sum_{l=1}^{d} c_r(2,l)\, y_r^l\, dr,$$

$$y_t^k = \int_{0}^{t} \sum_{l=1}^{d} c_r(k,l)\, y_r^l\, dr, \quad k \in \{3, \ldots, d\},$$

and by Gronwall's Lemma it follows that $y = 0$ on $[\tau_0, \tau_1)$. In order to prove $y = 0$ on $[\tau_\ell, \tau_{\ell+1})$ for some $\ell > 0$, note that we have $y_{\tau_\ell-} = 0$ by the induction assumption, so the claim follows again by Theorem 1.6 and Gronwall's Lemma.

□

1.4 Proof of the Main Result

1.4.1 Continuity w.r.t. the Initial Condition

The first step to prove Theorem 1.2 is to show the Lipschitz continuity of $x \mapsto (X_t(x))_t$ w.r.t. the sup-norm topology on a finite time interval:

Lemma 1.11. *For an arbitrary but fixed $T > 0$, let $(X_t(x))$ and $(X_t(y))$, $0 \leq t \leq T$, be solutions of (1.3) for some $x, y \in G$. Then, there exists a positive constant K, only depending on T, such that a.s.*

i) $\sup\limits_{t \in [0,T]} \|X_t(x) - X_t(y)\| \leq K\|x - y\|,$ *ii)* $\sup\limits_{t \in [0,T]} |l_t^i(x) - l_t^i(y)| \leq K\|x - y\|$ *for all i.*

Proof. By the assumption in (1.2), Theorem 2.1 in [28] ensures that Assumption 2.1 in [28] holds. Thus, we can apply Theorem 2.2 in [28] to obtain

$$\sup_{t \in [0,T]} \|X_t(x) - X_t(y)\| \leq K_1 \|x - y\| + K_1 \sup_{t \in [0,T]} \left\| \int_0^t [b(X_r(x)) - b(X_r(y))]\, dr \right\|,$$

and by the Lipschitz continuity of b we get

$$\sup_{t \in [0,T]} \|X_t(x) - X_t(y)\| \leq K_1 \|x - y\| + K_2 \int_0^T \sup_{r \leq s} \|X_r(x) - X_r(y)\|\, ds,$$

for some positive constants K_1 and K_2, and i) follows by the Gronwall Lemma. Using again Theorem 2.2 in [28], the Lipschitz property of b and i) we obtain ii). □

1.4.2 Computation of the Local Times

Recall the definition of the $\tilde{\mathbb{P}}(x)$ Brownian motion $W(x)$ in (1.4); the Skorohod SDE in (1.3) can be rewritten as follows:

$$X_t(x) = x + W_t(x) + \sum_i v_i\, l_t^i(x), \qquad t \geq 0, \tag{1.14}$$

so that

$$\langle X_t(x), n_i \rangle = \langle x, n_i \rangle + \langle W_t(x), n_i \rangle + \hat{L}_t^i(x) + l_t^i(x), \qquad t \geq 0,$$

since $< v_i, n_i > = 1$, where

$$\hat{L}_t^i(x) := \sum_{j \neq i} \langle v_j, n_i \rangle\, l_t^j(x), \qquad t \geq 0.$$

Note that $\langle W(x), n_i \rangle$ is again a Brownian motion under $\tilde{\mathbb{P}}(x)$ by Levy's characterization theorem, since n_i is a unit vector. The local time $l^i(x)$ is carried by the set of times t, when $\langle X_t(x), n_i \rangle - c_i = 0$, so that it can be computed directly by Skorohod's Lemma (see e.g. Lemma VI.2.1 in [56]). This yields

$$l_t^i(x) = \left[-\langle x, n_i \rangle + c_i - \inf_{s \leq t} \left(\langle W_s(x), n_i \rangle + \hat{L}_s^i(x) \right) \right]^+, \qquad t \geq 0.$$

Fix any q_n. Since $\langle X_{r_i(q_n)}(x), n_i \rangle - c_i = 0$ and $t \mapsto l_t^i(x)$ is increasing, we have for all $s \leq r_i(q_n)$:

$$\langle W_{r_i(q_n)}(x), n_i \rangle + \hat{L}_{r_i(q_n)}^i(x) = -\langle x, n_i \rangle + c_i - l_{r_i(q_n)}^i(x) \leq -\langle x, n_i \rangle + c_i - l_s^i(x)$$
$$= -\langle X_s(x), n_i \rangle + c_i + \langle W_s(x), n_i \rangle + \hat{L}_s^i(x) \tag{1.15}$$
$$\leq \langle W_s(x), n_i \rangle + \hat{L}_s^i(x).$$

Therefore, for all $t \in A_n$:

$$l_t^i(x) = l_{r_i(q_n)}^i(x) = \left[-\langle x, n_i \rangle + c_i - \langle W_{r_i(t)}(x), n_i \rangle - \hat{L}_{r_i(t)}^i(x) \right]^+. \tag{1.16}$$

Next we compute the local times of the process with perturbed starting point. Set $x_\varepsilon := x + \varepsilon v$, $\varepsilon \in \mathbb{R}$, $v \in \mathbb{R}^d$, where $|\varepsilon|$ is always supposed to be sufficiently small, such that x_ε lies in $G \setminus \bigcup_{i,j:\, i \neq j}(\partial G_i \cap \partial G_j)$. We start with a preparing lemma:

Lemma 1.12. *Let $i \in \{1, \ldots, N\}$ and $0 \leq s < t$ be arbitrary and let $\vartheta : \Omega \to [s, t]$ be the random variable such that a.s. $\langle W_\vartheta(x), n_i \rangle < \langle W_r(x), n_i \rangle$ for all $r \in [s,t] \setminus \{\vartheta\}$. Then, there exists a random $\Delta > 0$ such that a.s. ϑ is the only time, when $\langle W(x_\varepsilon), n_i \rangle = \langle W(x), n_i \rangle + \langle W(x_\varepsilon) - W(x), n_i \rangle$ attains its minimum over $[s,t]$ for all $|\varepsilon| < \Delta$.*

Proof. Since $\langle W(x), n_i \rangle$ is a Brownian motion under $\tilde{\mathbb{P}}(x)$, by Lemma 1 in [26] there exists a random variable γ such that every γ-Lipschitz perturbation of $\langle W(x), n_i \rangle$ attains its minimum only at ϑ. Using Lemma 1.11 and the Lipschitz continuity of b we find a $\Delta > 0$ such that $\sup_{r \in [s,t]} |\langle b(X_r(x_\varepsilon)) - b(X_r(x)), n_i \rangle| \leq \gamma$ for all $|\varepsilon| < \Delta$. This implies that $h(r) := \langle W_r(x_\varepsilon) - W_r(x), n_i \rangle = h(s) + \int_s^r \langle b(X_u(x_\varepsilon)) - b(X_u(x)), n_i \rangle\, du$ is a γ-Lipschitz perturbation for such ε, and the claim follows. □

Lemma 1.13. *For all i and q_n, $n \in \mathbb{N}$, there exists a random $\Delta_n^i > 0$ such that for all $|\varepsilon| < \Delta_n^i$ a.s.:*

$$l_{q_n}^i(x_\varepsilon) = \left[-\langle x_\varepsilon, n_i\rangle + c_i - \langle W_{r_i(q_n)}(x_\varepsilon), n_i\rangle - \hat{L}_{r_i(q_n)}^i(x_\varepsilon)\right]^+. \tag{1.17}$$

Proof. We need only to consider the case $s(q_n) = i$. Indeed, if $q_n < \inf C^i$ we can use Lemma 1.11 to find a $\Delta_n^i > 0$, such that $X_t(x_\varepsilon) \notin \partial G_i$ for all $t \in [0, q_n]$ and for all $|\varepsilon| < \Delta_n^i$, which implies $l_{q_n}^i(x_\varepsilon) = l_{q_n}^i(x) = 0$. If $q_n > \inf C^i$ and $s(t) \neq i$, we set $\tilde{q}_n := \sup\{q_k : q_k < q_n, s(q_k) = i\}$ and again by Lemma 1.11 there exists a $\Delta_n^i > 0$, such that $X_t(x_\varepsilon) \notin \partial G_i$ for all $t \in [\tilde{q}_n, q_n]$ and for all $|\varepsilon| < \Delta_n^i$, which implies $l_{q_n}^i(x_\varepsilon) = l_{\tilde{q}_n}^i(x_\varepsilon)$.

Let now q_n be such that $s(q_n) = i$. Using again Skorohod's Lemma, we obtain for all ε:

$$l_{q_n}^i(x_\varepsilon) = \left[-\langle x_\varepsilon, n_i\rangle + c_i - \inf_{s \leq q_n}\left(\langle W_s(x_\varepsilon), n_i\rangle + \hat{L}_s^i(x_\varepsilon)\right)\right]^+$$

$$= \left[-\langle x_\varepsilon, n_i\rangle + c_i - \inf_{s \leq q_n}(f_\varepsilon(s) + g_\varepsilon(s))\right]^+,$$

where $f_\varepsilon(s) := \langle W_s(x_\varepsilon), n_i\rangle + \hat{L}_s^i(x)$ and $g_\varepsilon(s) := \hat{L}_s^i(x_\varepsilon) - \hat{L}_s^i(x)$. From the calculation in (1.15) above we know that $\langle W(x), n_i\rangle + \hat{L}_s^i(x)$ attains its minimum over $[0, q_n]$ at $r_i(q_n)$, and we have to show that for sufficiently small $|\varepsilon|$:

$$\inf_{s \leq q_n}(f_\varepsilon(s) + g_\varepsilon(s)) = f_\varepsilon(r_i(q_n)) + g_\varepsilon(r_i(q_n)). \tag{1.18}$$

Recall that $q_n < \tau$, i.e. the process X hits the faces of the polyhedron G only successively, and recall that C^i is the support of $l^i(x)$. Thus, there exists a time $q_n^- < q_n$ such that $2d := l_{q_n}^i(x) - l_{q_n^-}^i(x) > 0$ and $X_s(x) \notin \bigcup_{j \neq i} \partial G_j$ for all $s \in [q_n^-, q_n]$ (note that we might have $q_n^- = 0$ in the case where $x \in \partial G_i$ and $q_n < \inf \bigcup_{j \neq i} C^j$). We apply Lemma 1.11 and find a $\Delta_n' > 0$ such that also $X_s(x_\varepsilon) \notin \bigcup_{j \neq i} \partial G_j$ for all $s \in [q_n^-, q_n]$ and $|\varepsilon| < \Delta_n'$. Hence, for such ε it follows that $\hat{L}^i(x), \hat{L}^i(x_\varepsilon)$ and g_ε are constant on $[q_n^-, q_n]$, so that $f_\varepsilon + g_\varepsilon$ attains its minimum over $[q_n^-, q_n]$ at the same time as $\langle W(x_\varepsilon), n_i\rangle$. By Lemma 1.12, possibly after choosing a smaller Δ_n', we know that this time is $r_i(q_n)$, so that

$$\inf_{s \in [q_n^-, q_n]}(f_\varepsilon(s) + g_\varepsilon(s)) = f_\varepsilon(r_i(q_n)) + g_\varepsilon(r_i(q_n)), \quad \forall |\varepsilon| < \Delta_n'. \tag{1.19}$$

Proceeding as in (1.15), we get for all $s \leq q_n^-$:

$$\langle W_{r_i(q_n)}(x), n_i\rangle + \hat{L}_{r_i(q_n)}^i(x) = -\langle x, n_i\rangle + c_i - l_{r_i(q_n)}^i(x) = -\langle x, n_i\rangle + c_i - l_{q_n}^i(x)$$
$$= -\langle x, n_i\rangle + c_i - l_{q_n^-}^i(x) - 2d \leq -\langle x, n_i\rangle + c_i - l_s^i(x) - 2d$$
$$= -\langle X_s(x), n_i\rangle + c_i + \langle W_s(x), n_i\rangle + \hat{L}_s^i(x) - 2d \tag{1.20}$$
$$\leq \langle W_s(x), n_i\rangle + \hat{L}_s^i(x) - 2d.$$

Using the Lipschitz continuity of b and Lemma 1.11, we find a $\Delta_n'' > 0$ such that

$$\sup_{s \leq q_n}|\langle W_s(x_\varepsilon) - W_s(x), n_i\rangle| = \sup_{s \leq q_n}\left|\int_0^s \langle b(X_r(x_\varepsilon)) - b(X_r(x)), n_i\rangle\, dr\right| \leq \frac{d}{2}, \quad \forall |\varepsilon| < \Delta_n'',$$

i.e. for such ε (1.20) implies

$$\inf_{s \leq q_n^-} f_\varepsilon(s) - f_\varepsilon(r_i(q_n)) = \inf_{s \leq q_n^-} \left(\langle W_s(x_\varepsilon), n_i \rangle + \hat{L}_s^i(x) \right) - \langle W_{r_i(q_n)}(x_\varepsilon), n_i \rangle - \hat{L}_{r_i(q_n)}^i(x)$$
$$\geq \inf_{s \leq q_n^-} \left(\langle W_s(x), n_i \rangle + \hat{L}_s^i(x) \right) - \sup_{s \leq q_n^-} |\langle W_s(x_\varepsilon) - W_s(x), n_i \rangle| \qquad (1.21)$$
$$- \langle W_{r_i(q_n)}(x_\varepsilon), n_i \rangle - \hat{L}_{r_i(q_n)}^i(x)$$
$$\geq \tfrac{3}{2} d - \langle W_{r_i(q_n)}(x_\varepsilon) - W_{r_i(q_n)}(x), n_i \rangle \geq d.$$

By Lemma 1.11 ii) there exists a random $\Delta_n''' > 0$ such that a.s.

$$\sup_{s \leq q_n} |g_\varepsilon(s)| < \frac{d}{2}, \qquad \forall |\varepsilon| < \Delta_n''. \qquad (1.22)$$

Now using (1.21) and (1.22) we obtain for $|\varepsilon| < \min(\Delta_n'', \Delta_n''')$:

$$\inf_{s \leq q_n^-} (f_\varepsilon(s) + g_\varepsilon(s)) \geq \inf_{s \leq q_n^-} f_\varepsilon(s) - \sup_{s \leq q_n^-} |g_\varepsilon(s)| > d + f_\varepsilon(r_i(q_n)) - \frac{d}{2}$$
$$= f_\varepsilon(r_i(q_n)) + \frac{d}{2} > f_\varepsilon(r_i(q_n)) + g_\varepsilon(r_i(q_n)), \qquad (1.23)$$

so that (1.18) follows from (1.19) and (1.23) for all $|\varepsilon| < \Delta_n^i := \min(\Delta_n', \Delta_n'', \Delta_n''')$. □

1.4.3 Computation of the Difference Quotients

Next we compute the difference quotients of X. For fixed $v \in \mathbb{R}^d$ we set $x_\varepsilon = x + \varepsilon v$, $\varepsilon \neq 0$ and

$$\eta_t(\varepsilon) := \frac{1}{\varepsilon} \left(X_t(x_\varepsilon) - X_t(x) \right), \qquad t \geq 0.$$

Let now $t \in [0, \tau)$ and n and ℓ be such that $t \in A_n$ and $t \in [\tau_\ell, \tau_{\ell+1})$. We choose $\Delta_n > 0$ such that a.s. for all $|\varepsilon| < \Delta_n$ we have for all i that $l_{q_n}^i(x) = l_{q_n}^i(x_\varepsilon) = 0$ if $q_n < \inf C^i$, and both of them are strictly positive if $q_n > \inf C^i$, and finally that formula (1.17) holds.
From (1.3) we deduce directly

$$X_t(x_\varepsilon) - X_t(x) = x_\varepsilon - x + W_t(x_\varepsilon) - W_t(x) + \sum_{j=1}^N v_j \left(l_t^j(x_\varepsilon) - l_t^j(x) \right). \qquad (1.24)$$

If $s(t) = 0$ we get immediately

$$\eta_t(\varepsilon) = v + \frac{1}{\varepsilon} \int_0^t (b(X_r(x_\varepsilon)) - b(X_r(x))) \, dr. \qquad (1.25)$$

Let us now consider the case $s(t) = i$, $i \in \{1, \ldots, N\}$. Then, possibly after choosing a smaller Δ_n, we may suppose that $l^j(x_\varepsilon)$ is constant on $[\tau_\ell, q_n \vee t]$ for every $j \neq i$ since $t < \tau$. In (1.24) we use (1.16) and (1.17) to obtain

$$X_t(x_\varepsilon) - X_t(x)$$
$$= x_\varepsilon - x + W_t(x_\varepsilon) - W_t(x) + \sum_{j \neq i} v_j \left(l^j_{r_i(q_n)}(x_\varepsilon) - l^j_{r_i(q_n)}(x) \right)$$
$$+ v_i \left(-\langle x_\varepsilon - x, n_i \rangle - \langle W_{r_i(q_n)}(x_\varepsilon) - W_{r_i(q_n)}(x), n_i \rangle - \left(\hat{L}^i_{r_i(q_n)}(x_\varepsilon) - \hat{L}^i_{r_i(q_n)}(x) \right) \right).$$

Hence,

$$\langle X_t(x_\varepsilon) - X_t(x), n_i \rangle$$
$$= \langle x_\varepsilon - x, n_i \rangle + \langle W_t(x_\varepsilon) - W_t(x), n_i \rangle + \hat{L}^i_{r_i(q_n)}(x_\varepsilon) - \hat{L}^i_{r_i(q_n)}(x)$$
$$- \langle x_\varepsilon - x, n_i \rangle - \langle W_{r_i(q_n)}(x_\varepsilon) - W_{r_i(q_n)}(x), n_i \rangle - \left(\hat{L}^i_{r_i(q_n)}(x_\varepsilon) - \hat{L}^i_{r_i(q_n)}(x) \right)$$
$$= \int_{r_i(q_n)}^t \langle b(X_r(x_\varepsilon)) - b(X_r(x)), n_i \rangle \, dr$$

and we get

$$\langle \eta_t(\varepsilon), n_i \rangle = \frac{1}{\varepsilon} \int_{r_i(q_n)}^t \langle b(X_r(x_\varepsilon)) - b(X_r(x)), n_i \rangle \, dr. \qquad (1.26)$$

Recall the definition of $v_i^\perp$ and $n_i^\perp$ in (1.1). Since $\langle v_i, n_i^\perp \rangle = -\langle v_i^\perp, n_i \rangle$, we have

$$\left\langle X_t(x_\varepsilon) - X_t(x), n_i^\perp \right\rangle$$
$$= \langle x_\varepsilon - x, n_i^\perp \rangle + \langle W_t(x_\varepsilon) - W_t(x), n_i^\perp \rangle + \sum_{j \neq i} \langle v_j, n_i^\perp \rangle \left(l^j_{r_i(q_n)}(x_\varepsilon) - l^j_{r_i(q_n)}(x) \right)$$
$$+ \langle v_i^\perp, n_i \rangle \left(\langle x_\varepsilon - x, n_i \rangle + \langle W_{r_i(q_n)}(x_\varepsilon) - W_{r_i(q_n)}(x), n_i \rangle + \hat{L}^i_{r_i(q_n)}(x_\varepsilon) - \hat{L}^i_{r_i(q_n)}(x) \right)$$
$$= \langle x_\varepsilon - x, n_i^\perp \rangle + \langle W_{r_i(q_n)}(x_\varepsilon) - W_{r_i(q_n)}(x), n_i^\perp \rangle + \sum_{j \neq i} \langle v_j, n_i^\perp \rangle \left(l^j_{r_i(q_n)}(x_\varepsilon) - l^j_{r_i(q_n)}(x) \right)$$
$$+ \left\langle x_\varepsilon - x, \langle v_i^\perp, n_i \rangle n_i \right\rangle + \left\langle W_{r_i(q_n)}(x_\varepsilon) - W_{r_i(q_n)}(x), \langle v_i^\perp, n_i \rangle n_i \right\rangle$$
$$+ \sum_{j \neq i} \left\langle v_j, \langle v_i^\perp, n_i \rangle n_i \right\rangle \left(l^j_{r_i(q_n)}(x_\varepsilon) - l^j_{r_i(q_n)}(x) \right) + \int_{r_i(q_n)}^t \left\langle b(X_r(x_\varepsilon)) - b(X_r(x)), n_i^\perp \right\rangle dr.$$

By the choice of $v_i^\perp$ and $n_i^\perp$, clearly $v_i^\perp = \langle v_i^\perp, n_i \rangle n_i + n_i^\perp$, so that

$$\left\langle X_t(x_\varepsilon) - X_t(x), n_i^\perp \right\rangle$$
$$= \left\langle x_\varepsilon - x + W_{r_i(q_n)}(x_\varepsilon) - W_{r_i(q_n)}(x) + \sum_{j \neq i} v_j \left(l^j_{r_i(q_n)}(x_\varepsilon) - l^j_{r_i(q_n)}(x) \right), v_i^\perp \right\rangle$$
$$+ \int_{r_i(q_n)}^t \left\langle b(X_r(x_\varepsilon)) - b(X_r(x)), n_i^\perp \right\rangle dr.$$

Since $W(x_\varepsilon) - W(x)$ is continuous, $\langle v_i, v_i^\perp \rangle = 0$ and $l^j(x_\varepsilon)$ and $l^j(x)$ are constant on $[\tau_\ell, r_i(q_n)]$ for all $j \neq i$ by the choice of Δ_n, we obtain

$$\left\langle \eta_t(\varepsilon), n_i^\perp \right\rangle = \left\langle \eta_{r_i(q_n)-}(\varepsilon), v_i^\perp \right\rangle + \frac{1}{\varepsilon} \int_{r_i(q_n)}^{t} \left\langle b(X_r(x_\varepsilon)) - b(X_r(x)), n_i^\perp \right\rangle dr. \tag{1.27}$$

and on the other hand we use again $v_i^\perp = \langle v_i^\perp, n_i \rangle n_i + n_i^\perp$ to obtain

$$\left\langle \eta_t(\varepsilon), n_i^\perp \right\rangle = \left\langle \eta_{\tau_\ell -}(\varepsilon), v_i^\perp \right\rangle + \frac{1}{\varepsilon} \int_{\tau_\ell}^{t} \left\langle b(X_r(x_\varepsilon)) - b(X_r(x)), n_i^\perp \right\rangle dr$$
$$+ \langle v_i^\perp, n_i \rangle \frac{1}{\varepsilon} \int_{\tau_\ell}^{r_i(q_n)} \langle b(X_r(x_\varepsilon)) - b(X_r(x)), n_i \rangle dr. \tag{1.28}$$

Finally, for every $k \in \{3, \ldots, d\}$, we have $\langle v_i, n_i^k \rangle = 0$, so that

$$\left\langle X_t(x_\varepsilon) - X_t(x), n_i^k \right\rangle = \left\langle x_\varepsilon - x + W_t(x_\varepsilon) - W_t(x) + \sum_{j \neq i} v_j \left(l^j_{r_i(q_n)}(x_\varepsilon) - l^j_{r_i(q_n)}(x) \right), n_i^k \right\rangle.$$

Thus,

$$\left\langle \eta_t(\varepsilon), n_i^k \right\rangle = \left\langle \eta_{r_i(q_n)-}(\varepsilon), n_i^k \right\rangle + \frac{1}{\varepsilon} \int_{r_i(q_n)}^{t} \left\langle b(X_r(x_\varepsilon)) - b(X_r(x)), n_i^k \right\rangle dr \tag{1.29}$$

and

$$\left\langle \eta_t(\varepsilon), n_i^k \right\rangle = \left\langle \eta_{\tau_\ell -}(\varepsilon), n_i^k \right\rangle + \frac{1}{\varepsilon} \int_{\tau_\ell}^{t} \left\langle b(X_r(x_\varepsilon)) - b(X_r(x)), n_i^k \right\rangle dr. \tag{1.30}$$

Since $\{n_i^k; k = 1, \ldots, d\}$ is an orthonormal basis of $\mathbb{R}^d$ we obtain by (1.26), (1.27) and (1.29),

$$\eta_t(\varepsilon) = \langle \eta_t(\varepsilon), n_i \rangle n_i + \left\langle \eta_t(\varepsilon), n_i^\perp \right\rangle n_i^\perp + \sum_{k=3}^{d} \left\langle \eta_t(\varepsilon), n_i^k \right\rangle n_i^k$$
$$= \langle \eta_{r_i(t)-}(\varepsilon), v_i^\perp \rangle n_i^\perp + \sum_{k=3}^{d} \langle \eta_{r_i(t)-}(\varepsilon), n_i^k \rangle n_i^k + \frac{1}{\varepsilon} \int_{r_i(t)}^{t} (b(X_r(x_\varepsilon)) - b(X_r(x))) \, dr. \tag{1.31}$$

Then, from (1.25) and (1.31) we get

$$\eta_t(\varepsilon) = \eta_{a_n}(\varepsilon) + \frac{1}{\varepsilon} \int_{a_n}^{t} [b(X_r(x_\varepsilon)) - b(X_r(x))] \, dr$$
$$= \eta_{a_n}(\varepsilon) + \int_{a_n}^{t} \sum_{k=1}^{d} \left[\int_0^1 \frac{\partial b}{\partial x^k}(X_r^{\alpha,\varepsilon}) \, d\alpha \right] \eta_r^{kj}(\varepsilon) \, dr$$
$$= \eta_{a_n}(\varepsilon) + \int_{a_n}^{t} \int_0^1 Db(X_r^{\alpha,\varepsilon}) \cdot \eta_r(\varepsilon) \, d\alpha \, dr, \quad t \in A_n,$$

where $X_r^{\alpha,\varepsilon} := \alpha X_r(x_\varepsilon) + (1-\alpha) X_r(x)$, $\alpha \in [0,1]$.

1.4.4 Proof of the Differentiability

Now we argue similar to Step 5 in the proof of Theorem 1 in [26]. Recall that $\sup_{s\in[0,t]} \|\eta_s(\varepsilon)\| \le c_1$ for some positive constant c_1 and $\varepsilon \ne 0$ by Lemma 1.11. Let $(\varepsilon_\nu)_\nu$ be any sequence converging to zero. By a diagonal procedure, we can extract a subsequence $(\nu_l)_l$ such that $\eta_{r(q_n)}(\varepsilon_{\nu_l})$ has a limit $\hat\eta_{r(q_n)} \in \mathbb{R}^d$ as $l \to \infty$ for all $n \in \mathbb{N}$. Let now $\hat\eta : [0,\tau)\backslash C \to \mathbb{R}^d$ be the unique solution of

$$\hat\eta_t := \hat\eta_{r(q_n)} + \int_{r(q_n)}^t Db(X_r(x)) \cdot \hat\eta_r \, dr, \qquad t \in A_n.$$

Then, we get for $|\varepsilon| \in (0, \Delta_n)$ and $t \in A_n$,

$$\|\eta_t(\varepsilon) - \hat\eta_t\| \le \|\eta_{r(q_n)}(\varepsilon) - \hat\eta_{r(q_n)}\| + c_1 \sup_{r \in A_n} \|Db(X_r^{\alpha,\varepsilon}) - Db(X_r(x))\|$$

$$+ c_2 \int_0^t \|\eta_r(\varepsilon) - \hat\eta_r\| \, dr,$$

$c_2 > 0$ denoting the Lipschitz constant of b. Since $\eta_{r(q_n)}(\varepsilon_{\nu_l}) \to \hat\eta_{r(q_n)}$, $X^{\alpha,\varepsilon_{\nu_l}} \to X_r(x)$ uniformly in $r \in [0,t]$ and the derivatives of b are continuous, we obtain by Gronwall's Lemma that $\eta_t(\varepsilon_{\nu_l})$ converges to $\hat\eta_t$ uniformly in $t \in A_n$ for every n. Thus, since C has zero Lebesgue measure, $\eta_t(\varepsilon_{\nu_l})$ converges to $\hat\eta_t$ for all $t \in [0,\tau)\backslash C$ as $l \to \infty$ and by the dominated convergence theorem we get

$$\hat\eta_t = v + \int_0^t Db(X_r(x)) \cdot \hat\eta_r \, dr, \qquad t \in [0, \inf C),$$

$$\hat\eta_t = \hat\eta_{r(t)} + \int_{r(t)}^t Db(X_r(x)) \cdot \hat\eta_r \, dr, \qquad t \in [\inf C, \tau),$$

where we have extended $\hat\eta$ on C such that it is right-continuous. Since C has zero Lebesgue measure this does not affect the value of $\hat\eta_t$, $t \in [0,\tau)\backslash C$. The proof of Theorem 1.2 is now complete once we have shown that $\hat\eta$ does not depend on the chosen subsequence. To that aim we define $y_t^\varepsilon = (y_t^{1,\varepsilon}, \ldots, y_t^{d,\varepsilon})$, $t \ge 0$, by

$$y_t^{k,\varepsilon} := \begin{cases} \langle \eta_t(\varepsilon), e^k \rangle & \text{if } s(t) = 0, \\ \langle \eta_t(\varepsilon), n_i^k \rangle & \text{if } s(t) = i. \end{cases}$$

Furthermore, we set

$$c_t^\varepsilon(k,l) := \begin{cases} \int_0^1 \langle e^k, Db(X_t^{\alpha,\varepsilon}) \cdot e^l \rangle \, d\alpha & \text{if } s(t) = 0, \\ \int_0^1 \langle n_i^k, Db(X_t^{\alpha,\varepsilon}) \cdot n_i^l \rangle \, d\alpha & \text{if } s(t) = i. \end{cases}$$

Then, since

$$\frac{1}{\varepsilon}[b(X_t(x_\varepsilon)) - b(X_t(x))] = \sum_{l=1}^d \left[\int_0^1 D_{n_i^l} b(X_t^{\alpha,\varepsilon}) \, d\alpha \right] \langle \eta_t(\varepsilon), n_i^l \rangle,$$

we obtain on one hand from (1.25) that for $t < \inf C$ and ε small enough,

$$y_t^{k,\varepsilon} = \langle v, e^k \rangle + \int_0^t \sum_{l=1}^d c_r^\varepsilon(k,l) \, y_r^{l,\varepsilon} \, dr, \qquad k \in \{1, \ldots, d\}.$$

1.4 Proof of the Main Result

On the other hand from (1.26), (1.28) and (1.30) we get for sufficiently small ε that y^ε satisfies for $t \in [\tau_\ell, \tau_{\ell+1})$ and i such that $s(t) = i$:

$$y_t^{1,\varepsilon} = \int_{r_i(t)}^t \sum_{l=1}^d c_r^\varepsilon(1,l)\, y_r^{l,\varepsilon}\, dr,$$

$$y_t^{2,\varepsilon} = \langle v_i^\perp, n_i\rangle\, y_{\tau_\ell-}^{1,\varepsilon} + y_{\tau_\ell-}^{2,\varepsilon} + \int_{\tau_\ell}^t \sum_{l=1}^d c_r^\varepsilon(2,l)\, y_r^{l,\varepsilon}\, dr + \langle v_i^\perp, n_i\rangle \int_{\tau_\ell}^{r_i(t)} \sum_{l=1}^d c_r^\varepsilon(2,l)\, y_r^{l,\varepsilon}\, dr,$$

$$y_t^{k,\varepsilon} = y_{\tau_\ell-}^{k,\varepsilon} + \int_{\tau_\ell}^t \sum_{l=1}^d c_r^\varepsilon(k,l)\, y_r^{l,\varepsilon}\, dr, \qquad k \in \{3,\ldots,d\}.$$

We set

$$y_t^k := \begin{cases} \langle \hat\eta_t, e^k\rangle & \text{if } s(t) = 0, \\ \langle \hat\eta_t, n_i^k\rangle & \text{if } s(t) = i, \end{cases} \quad \text{and} \quad c_t(k,l) := \begin{cases} \langle e^k, Db(X_t(x)) \cdot e^l\rangle & \text{if } s(t) = 0, \\ \langle n_i^k, Db(X_t(x)) \cdot n_i^l\rangle & \text{if } s(t) = i. \end{cases}$$

Since $\eta_t(\varepsilon)$ converges along the chosen subsequence to $\hat\eta_t$ uniformly in $t \in A_n$ for every n, we obtain by the dominated convergence theorem that y_t satisfies the system (1.9). In order to complete the proof of Theorem 1.2 we shall now prove that $\hat\eta$ is characterized by the system (1.9).

Lemma 1.14. *The system* (1.9) *admits a unique solution.*

Proof. We shall prove uniqueness of the solution on every interval $[\tau_\ell, \tau_{\ell+1})$ by induction over ℓ. If $\inf C > 0$, on $[\tau_0, \tau_1) = [0, \inf C)$ existence and uniqueness is clear. Otherwise, the initial condition is specified as in Theorem 1.9 and we get existence and uniqueness on $[\tau_0, \tau_1)$ by the same argument as for general ℓ. For $t \in [\tau_\ell, \tau_{\ell+1})$, $\ell \geq 0$, the system is given by

$$y_t^1 = y_{\tau_\ell}^1 + \int_{r_i(t)}^t \sum_{l=1}^d c_r(1,l)\, y_r^l\, dr,$$

$$y_t^2 = y_{\tau_\ell}^2 + \int_{\tau_\ell}^t \sum_{l=1}^d c_r(2,l)\, y_r^l\, dr + \langle v_i^\perp, n_i\rangle \int_{\tau_\ell}^{r_i(t)} \sum_{l=1}^d c_r(2,l)\, y_r^l\, dr,$$

$$y_t^k = y_{\tau_\ell}^k + \int_{\tau_\ell}^t \sum_{l=1}^d c_r(k,l)\, y_r^l\, dr, \qquad k \in \{3,\ldots,d\},$$

where the initial condition is uniquely specified by the induction assumption:

$$y_{\tau_\ell}^1 = 0, \quad y_{\tau_\ell}^k = \langle v_i^\perp, n_i\rangle\, y_{\tau_\ell-}^1 + y_{\tau_\ell-}^2, \quad y_{\tau_\ell}^k = y_{\tau_\ell-}^k.$$

For any fixed $T > 0$, let H be the totality of $\mathbb{R}^d$-valued adapted processes (φ_t), $t \in [0,T]$, whose paths are a.s. càdlàg and which satisfy $\sup_{t \in [0,T]} \mathbb{E}[\|\varphi_t\|^2] < \infty$. On H we introduce the norm

$$\|\varphi\|_H = \sup_{t \in [0,T]} \mathbb{E}\left[\|\varphi_t\|^2\right]^{1/2},$$

and for any $\varphi \in H$ we define the process $I(\varphi)$ by

$$I(\varphi)_t^1 = \mathbb{1}_{\{t \in [\tau_\ell, \tau_{\ell+1})\}} \left(y_{\tau_\ell}^1 + \int_{r_i(t)}^t \sum_{l=1}^d c_r(1,l) \, \varphi_r^l \, dr \right),$$

$$I(\varphi)_t^2 = \mathbb{1}_{\{t \in [\tau_\ell, \tau_{\ell+1})\}} \left(y_{\tau_\ell}^2 + \int_{\tau_\ell}^t \sum_{l=1}^d c_r(2,l) \, \varphi_r^l \, dr + \langle v_i^\perp, n_i \rangle \int_{\tau_\ell}^{r_i(t)} \sum_{l=1}^d c_r(2,l) \, \varphi_r^l \, dr \right),$$

$$I(\varphi)_t^k = \mathbb{1}_{\{t \in [\tau_\ell, \tau_{\ell+1})\}} \left(y_{\tau_\ell}^k + \int_{\tau_\ell}^t \sum_{l=1}^d c_r(k,l) \, \varphi_r^l \, dr \right), \qquad k \in \{3, \ldots, d\},$$

for $t \in [0, T]$. Since the system is linear in y with uniformly bounded coefficients, one can easily verify that $I(\varphi) \in H$ for every $\varphi \in H$ and that for any $\varphi, \psi \in H$

$$\|I(\varphi) - I(\psi)\|_H \leq c \|\varphi - \psi\|_H,$$

for some constant c not depending on φ and ψ. Hence, we obtain uniqueness on $[\tau_\ell, \tau_{\ell+1} \wedge T)$ by standard arguments via Picard-iteration. Since T is arbitrary, we get uniqueness on $[\tau_\ell, \tau_{\ell+1})$. □

Chapter 2
SDEs in a Smooth Domain with normal Reflection

2.1 Introduction

This chapter deals with the pathwise differentiability for the solution $(X_t(x))_{t\geq 0}$ of a stochastic differential equation (SDE) of the Skorohod type in a smooth bounded domain $G \subset \mathbb{R}^d$, $d \geq 2$, with normal reflection at the boundary. Again, the process $(X_t(x))$ is driven by a d-dimensional standard Brownian motion and by a drift term, whose coefficients are supposed to be continuously differentiable and Lipschitz continuous, i.e. existence and uniqueness of the solution are ensured by the results of Lions and Sznitman in [51].

We prove that for every $t > 0$ the solution $X_t(x)$ is differentiable w.r.t. the deterministic initial value x and we give a representation of the derivatives in terms of an ordinary differential equation. As an easy side result, we provide a Bismut-Elworthy formula for the gradient of the transition semigroup.

The resulting derivatives evolve according to a simple linear ordinary differential equation, when the process is away from the boundary, and they have a discontinuity and are projected to the tangent space, when the process hits the boundary. This evolution becomes rather complicated because of the structure of the set of times, when the process hits the boundary, which is known to be a.s. a closed set with zero Lebesgue measure without isolated points. The system is similar to the one introduced by Airault in [1] in order to develop probabilistic representations for the solutions of linear PDE systems with mixed Dirichlet-Neumann conditions in a smooth domain in $\mathbb{R}^d$. A further similar system appears in Section V.6 in [42], which deals with the heat equation for diffusion processes on manifolds with boundary conditions. In a sense, our result can be considered as a pathwise version of the results in [1, 42].

Again we shall use some of the techniques established in [26], where Deuschel and Zambotti proved such a pathwise differentiability result with respect to the initial data for diffusion processes in the domain $G = [0, \infty)^d$. The proof of the main result in [26] is based on the fact that a Brownian path, which is perturbed by adding a Lipschitz path with a sufficiently small Lipschitz constant, attains its minimum at the same time as the original path (see Lemma 1 in [26]). This is due to the fact that a Brownian path leaves its minimum faster than linearly. In [26] as well as in the previous chapter this is used in order to provide an exact computation of the reflection term in the difference quotient via Skorohod's lemma.

In this chapter, the approach is quite similar: Using localization techniques introduced by Anderson and Orey (cf. [5]) we transform the SDE locally into an SDE on a halfspace (cf. Section 2.2.4 below). Then, in order to compute the local time we need to deal with the pathwise minimum of a continuous martingale in place of the standard Brownian motion. Since the perturbations are now no longer Lipschitz continuous, i.e. Lemma 1 of [26] does not apply, and because of the asymptotics of a Brownian path around its minimum (cf. Lemma 2.12 below) one cannot necessarily expect that an analogous statement to Lemma 1 in [26] holds true in this case. Nevertheless, one can show that the minimum times converge sufficiently fast to obtain differentiability (see Proposition 2.13).

Another crucial ingredient in the proof is the Lipschitz continuity of the solution with respect to the initial data. This was proven by Burdzy, Chen and Jones in Lemma 3.8 in [16] for the reflected Brownian motion without drift in planar domains, but the arguments can easily be transferred into our setting (see Proposition 2.8). This will give pathwise convergence of the difference quotients along a subsequence. In order to show that the limit does not depend on the chosen subsequence, we shall characterize the limit in coordinates with respect to a moving frame as the solution of an SDE-like equation (cf. Section 4 in [1]).

A pathwise differentiability result w.r.t. the initial position of a reflected Brownian motion in smooth domains has also been proven by Burdzy in [14] using excursion theory. The resulting derivative is characterized as a linear map represented by a multiplicative functional for reflected Brownian motion, which has been introduced in Theorem 3.2 of [17]. In contrast to our main results, the SDE considered in [14] does not contain a drift term and the differentiability is shown for the trace process, while we consider the process on the original time-scale. However, we can recover the term, which is mainly characterizing the derivative in [14], describing the influence of curvature of ∂G (cf. Remark 2.6 below).

In another recent article [54] Pilipenko studies flow properties for SDEs with reflection and receives Sobolev differentiability in the initial value. In general, reflected Brownian motions have been investigated in several articles, where the question of coalescence or noncoalescence of synchronous couplings is of particular interest. For planar convex domains this has been studied by Cranston and Le Jan in [22] and [23], for some classes of non-smooth domains by Burdzy and Chen in [15], and for two-dimensional smooth domains by Burdzy, Chen and Jones in [16], while the case of a multi-dimensional smooth domain is still an open problem.

The material presented in this chapter is contained in [7]. The chapter is organized as follows: In Section 2.2 we give the precise setup and some further preliminaries and we present the main results. Section 2.3 is devoted to the proof of the main results.

2.2 Main Results and Preliminaries

2.2.1 General Notation

As in the last chapter we denote by $\|.\|$ the Euclidian norm, by $\langle .,. \rangle$ the canonical scalar product and by $e = (e^1, \ldots, e^d)$ the standard basis in $\mathbb{R}^d$, $d \geq 2$. Let $G \subset \mathbb{R}^d$ be a connected closed bounded domain with C^3-smooth boundary and G_0 its interior and let $n(x)$, $x \in \partial G$, denote the inner normal field. For any $x \in \partial G$, let

$$\pi_x(z) := z - \langle z, n(x) \rangle n(x), \quad z \in \mathbb{R}^d,$$

denote the orthogonal projection onto the tangent space. The closed ball in $\mathbb{R}^d$ with center x and radius r will be denoted by $B_r(x)$. The transposition of a vector $v \in \mathbb{R}^d$ and of a matrix $A \in \mathbb{R}^{d \times d}$ will be denoted by v^* and A^*, respectively. The set of continuous real-valued functions on G is denoted by $C(G)$, and $C_b(G)$ denotes the set of those functions in $C(G)$ that are bounded on G. For each $k \in \mathbb{N}$, $C^k(G)$ denotes the set of real-valued functions that are k-times continuously differentiable in G, and $C_b^k(G)$ denotes the set of those functions in $C^k(G)$ that are bounded and have bounded partial derivatives up to order k. Furthermore, for $f \in C^1(G)$ we denote by ∇ the gradient of f and in the case where f is $\mathbb{R}^d$-valued by Df the Jacobi matrix. Finally, Δ denotes the Laplace differential operator on $C^2(G)$ and $D_v := \langle v, \nabla \rangle$ the directional derivative operator associated with the direction $v \in \mathbb{R}^d$. The symbols c and c_i, $i \in \mathbb{N}$, will denote constants, whose value may only depend on some quantities specified in the particular context.

2.2.2 Skorohod SDE

For any starting point $x \in G$, we consider the following stochastic differential equation of the Skorohod type:

$$\begin{aligned} X_t(x) &= x + \int_0^t b(X_r(x)) \, dr + w_t + \int_0^t n(X_r(x)) \, dl_r(x), \quad t \geq 0, \\ X_t(x) &\in G, \quad dl_t(x) \geq 0, \quad \int_0^\infty \mathbb{1}_{G_0}(X_t(x)) \, dl_t(x) = 0, \quad t \geq 0, \end{aligned} \quad (2.1)$$

where w is a d-dimensional Brownian motion on a complete probability space $(\Omega, \mathcal{F}, \mathbb{P})$ and $l(x)$ denotes the local time of $X(x)$ in ∂G, i.e. it increases only at those times, when $X(x)$ is at the boundary of G. The components $b^i : G \to \mathbb{R}$ of b are supposed to be in $C^1(G)$ and Lipschitz continuous. Then, existence and uniqueness of strong solutions of (2.1) are guaranteed by the results in [60] in the case, where G is a convex set, and for arbitrary smooth G by the results in [51]. The local time l is carried by the set

$$C := \{ s \geq 0 : X_s(x) \in \partial G \}.$$

We define

$$r(t) := \sup(C \cap [0, t])$$

with the convention $\sup \emptyset := 0$. Then, C is known to be a.s. a closed set of zero Lebesgue measure without isolated points and $t \mapsto r(t)$ is locally constant and right-continuous. Let $(A_n)_n$ be the family of connected components of the complement of C. A_n is open, so that there exists $q_n \in A_n \cap \mathbb{Q}$, $n \in \mathbb{N}$. Note that for each $t > \inf C$ we have $X_{r(t)}(x) \in \partial G$.

For later use we introduce now a moving frame. Let $x \mapsto O(x)$ be a mapping on G taking values in the space of orthogonal matrices, which is twice continuouslsy differentiable, such that for $x \in \partial G$ the first row of $O(x)$ coincides with $n(x)$. Such a mapping, which exists locally, can be constructed on the whole domain G by using the partition of unity. Writing $O_s = O(X_s(x))$ we get by Itô's formula

$$dO_s = \sum_{k=1}^{d} \alpha_k(X_s(x))\, dw_s^k + \beta(X_s(x))\, ds + \gamma(X_s(x))\, dl_s(x), \qquad (2.2)$$

for some coefficient functions α_k, β and γ.

2.2.3 Main Results

Theorem 2.1. *For all $t > 0$ and $x \in G$ a.s. the mapping $y \mapsto X_t(y)$ is differentiable at x and, setting $\eta_t := D_v X_t(x) = \lim_{\varepsilon \to 0} (X_t(x + \varepsilon v) - X_t(x))/\varepsilon$, $v \in \mathbb{R}^d$, there exists a right-continuous modification of η such that a.s. for all $t > 0$:*

$$\begin{aligned}
\eta_t &= v + \int_0^t Db(X_r(x)) \cdot \eta_r\, dr, & \text{if } t < \inf C, \\
\eta_t &= \pi_{X_{r(t)}(x)}(\eta_{r(t)-}) + \int_{r(t)}^t Db(X_r(x)) \cdot \eta_r\, dr, & \text{if } t \geq \inf C.
\end{aligned} \qquad (2.3)$$

Remark 2.2. If $x \in \partial G$, $t = 0$ is a.s. an accumulation point of C and we have $r(t) > 0$ a.s. for every $t > 0$. Therefore, in that case $\eta_0 = v$ and $\eta_{0+} = \pi_x(v)$, i.e. there is discontinuity at $t = 0$.

Remark 2.3. The equation (2.3) does not characterize the derivatives, since it does not admit a unique solution. Indeed, if the process (η_t) solves (2.3), then the process $(1 + l_t(x))\eta_t$, $t \geq 0$, also does. A characterizing equation for the derivatives is given Theorem 2.5 below.

Note that this result corresponds to that for the domain $G = [0, \infty)^d$ in Theorem 1 in [26]. The proof of Theorem 2.1 as well as the proofs of Theorem 2.5 and Corollary 2.7 below are postponed to Section 2.3. As soon as pathwise differentiability is established, we can immediately provide a Bismut-Elworthy formula: Define for all $f \in C_b(G)$ the transition semigroup $P_t f(x) := \mathbb{E}[f(X_t(x))]$, $x \in G$, $t > 0$, associated with X.

Corollary 2.4. *Setting $\eta_t^{ij} := \partial X_t^i(x)/\partial x^j$, we have for all $f \in C_b(G)$, $t > 0$ and $x \in G$:*

$$\frac{\partial}{\partial x^i} P_t f(x) = \frac{1}{t}\, \mathbb{E}\left[f(X_t(x)) \int_0^t \sum_{k=1}^d \eta_r^{ki}\, dw_r^k \right], \qquad i \in \{1, \ldots, d\}, \qquad (2.4)$$

and if $f \in C_b^1(G)$:

$$\frac{\partial}{\partial x^i} P_t f(x) = \sum_{k=1}^d \mathbb{E}\left[\frac{\partial f}{\partial x^k}(X_t(x))\, \eta_t^{ki} \right], \qquad i \in \{1, \ldots, d\}. \qquad (2.5)$$

Proof. Formula (2.5) is straightforward from the differentiability statement in Theorem 2.1 and the chain rule. For formula (2.4) see the proof of Theorem 2 in [26]. □

2.2 Main Results and Preliminaries

From the representation of the derivatives in (2.3) it is obvious that $(\eta_t)_t$ evolves according to a linear differential equation, when the process X is in the interior of G, and that it is projected to the tangent space, when X hits the boundary. Furthermore, if X is at the boundary at some time t_0 and we have $r(t_0-) \neq r(t_0)$, i.e. t_0 is the endpoint of an excursion interval, then also η has a discontinuity at t_0 and jumps as follows:

$$\eta_{t_0} = \pi_{X_{t_0}(x)}(\eta_{t_0-}). \tag{2.6}$$

Consequently, we observe that at each time t_0 as above, η is projected to the tangent space and jumps in direction of $n(X_{t_0}(x))$ or $-n(X_{t_0}(x))$, respectively. Finally, if $X_{t_0}(x) \in \partial G$ and $t \mapsto r(t)$ is continuous in $t = t_0$, there is also a projection of η, but since in this case η_{t_0-} is in the tangent space, the projection has no effect and η is continuous at time t_0.

Set $Y_t := O_t \cdot \eta_t$, $t \geq 0$, where O_t denotes the moving frame introduced in Section 2.2.2. Let $P = \operatorname{diag}(e^1)$ and $Q = \operatorname{Id} - P$ and

$$Y_t^1 = P \cdot Y_t \quad \text{and} \quad Y_t^2 = Q \cdot Y_t$$

to decompose the space $\mathbb{R}^d$ into the direct sum $\operatorname{Im} P \oplus \operatorname{Ker} P$. We define the coefficient functions $c(t)$ and $d(t)$ to be such that

$$\sum_{k=1}^d \begin{pmatrix} c_k^1(t) & c_k^2(t) \\ c_k^3(t) & c_k^4(t) \end{pmatrix} dw_t^k + \begin{pmatrix} d^1(t) & d^2(t) \\ d^3(t) & d^4(t) \end{pmatrix} dt$$

$$= \sum_{k=1}^d \alpha_k(X_t(x)) \cdot O_t^{-1} \, dw_t^k + \left[O_t \cdot Db(X_t(x)) \cdot O_t^{-1} + \beta(X_t(x)) \cdot O_t^{-1} \right] dt.$$

Furthermore, we set $\gamma^2(t) := \gamma(X_t(x)) \cdot O_t^{-1} \cdot Q$.

Theorem 2.5. *There exists a right-continuous modification of η and Y, respectively, such that Y is characterized as the unique solution of*

$$Y_t^1 = \mathbb{1}_{\{t < \inf C\}} \left(Y_0^1 + \sum_{k=1}^d \int_0^t \left(c_k^1(s) Y_s^1 + c_k^2(s) Y_s^2 \right) dw_s^k + \int_0^t \left(d^1(s) Y_s^1 + d^2(s) Y_s^2 \right) ds \right)$$

$$+ \mathbb{1}_{\{t \geq \inf C\}} \left(\sum_{k=1}^d \int_{r(t)}^t \left(c_k^1(s) Y_s^1 + c_k^2(s) Y_s^2 \right) dw_s^k + \int_{r(t)}^t \left(d^1(s) Y_s^1 + d^2(s) Y_s^2 \right) ds \right)$$

$$Y_t^2 = Y_0^2 + \sum_{k=1}^d \int_0^t \left(c_k^3(s) Y_s^1 + c_k^4(s) Y_s^2 \right) dw_s^k + \int_0^t \left(d^3(s) Y_s^1 + d^4(s) Y_s^2 \right) ds$$

$$+ \int_0^t \left(\Phi_s^2 + \gamma^2(s) \right) Y_s^2 \, dl_s(x),$$

where

$$\Phi_t^2 := Q \cdot O_t \cdot Dn(X_t(x)) \cdot O_t^{-1} \cdot Q, \quad t \in C = \operatorname{supp} dl(x),$$

with the initial condition $Y_0^1 = P \cdot O(x) \cdot v$ and $Y_0^2 = Q \cdot O(x) \cdot v$.

Remark 2.6. Note that for all $t \in C = \operatorname{supp} dl(x)$,
$$\Phi_t^2 := Q \cdot O_t \cdot Dn(X_t(x)) \cdot O_t^{-1} \cdot Q = -Q \cdot O_t \cdot S(X_t(x)) \cdot O_t^{-1} \cdot Q,$$
where for every $x \in \partial G$, $S(x)$ denotes the symmetric linear endomorphism acting on the tangent space at x, which is known as the shape operator or the Weingarten map, characterized by the relation $S(x)v = -D_v n(x)$ for all v in the tangent space at x. The eigenvalues of $S(x)$ are the principal curvatures of ∂G at x, and its determinant is the Gaussian curvature. Hence, the linear term in the equation for the derivatives in [14] can be recovered in our results.

Finally, we give another confirmation of the results, namely they will imply that the Neumann condition holds for X.

Corollary 2.7. *For all $f \in C_b(G)$ and $t > 0$, the transition semigroup $P_t f(x) := \mathbb{E}[f(X_t(x))]$, $x \in G$, satisfies the Neumann condition at ∂G:*
$$x \in \partial G \Longrightarrow D_{n(x)} P_t f(x) = 0.$$

2.2.4 Localization

In order to prove Theorem 2.1 we shall use the localization technique introduced in [5]. Let $\{U_0, U_1, \ldots\}$ be a countable or finite family of relatively open subsets of G covering G_0. Every U_m is attached with a coordinate system, i.e. with a mapping $u_m : U_m \to \mathbb{R}^d$, giving each point $x \in U_m$ the coordinates $u_m(x) = (u_m^1(x), \ldots, u_m^d(x))$ such that:

i) $U_0 \subseteq G_0$ and the corresponding coordinates are the original Euclidian coordinates. If $m > 0$ the mapping u_m is one to one and twice continuously differentiable and we have
$$U_m \cap \partial G = \{x \in U_m : u_m^1(x) = 0\}, \qquad U_m \cap G_0 = \{x \in U_m : u_m^1(x) > 0\}.$$

ii) There is a positive constant d_0 such that for every $x \in G$ there exists an index $m(x) \in \mathbb{N}$ such that $B_{d_0}(x) \subseteq U_{m(x)}$.

iii) For every $m > 0$, $\langle \nabla u_m^i(x), n(x) \rangle = \delta_{1i}$ for all $x \in U_m \cap \partial G$.

iv) For every $m \geq 0$ and $i \in \{1, \ldots, d\}$, the functions
$$b_m^i : U_m \to \mathbb{R} \quad x \mapsto \langle \nabla u_m^i(x), b(x) \rangle + \tfrac{1}{2} \Delta u_m^i(x),$$
$$\sigma_m^i : U_m \to \mathbb{R}^d \quad x \mapsto \nabla u_m^i(x),$$
satisfy
$$\sup_m \sup_{x \in U_m} \left(|b_m^i(x)| + \|\sigma_m^i(x)\| \right) < \infty,$$
and there exists a constant c, not depending on m and i, such that
$$|b_m^i(x) - b_m^i(y)| + \|\sigma_m^i(x) - \sigma_m^i(y)\| \leq c \|x - y\|, \qquad \forall x, y \in U_m.$$

Note that these conditions imply $n(x) = \nabla u_m^1(x)$ for all $x \in U_m \cap \partial G$. Since ∂G is supposed to be C^3, the function b_m^i and σ_m^i are continuously differentiable. Fix any $\delta_0 \in (0, d_0)$. We define a sequence of stopping times $(\tau_\ell)_\ell$ by

$$\tau_0 := 0, \qquad \tau_{\ell+1} := \{t > \tau_\ell : \operatorname{dist}(X_t(x), \partial U_{m_\ell}) < \delta_0\}, \quad \ell \geq 0,$$

where, for every $\ell \geq 0$, $m_\ell := m(X_{\tau_\ell}(x)) \in \mathbb{N}$ such that $B_{d_0}(X_{\tau_\ell}(x)) \subseteq U_{m_\ell}$. The dependence of τ_ℓ on x will be suppressed in the notation. Note that by construction $X_t(x) \in U_{m_\ell}$ for all $t \in [\tau_\ell, \tau_{\ell+1}]$ and $m_\ell \neq m_{\ell+1}$ for every ℓ since $\delta_0 < d_0$.

Using Itô's formula we get for $t \in [\tau_\ell, \tau_{\ell+1}]$:

$$\begin{aligned}
u_{m_\ell}(X_t(x)) =& u_{m_\ell}(X_{\tau_\ell}(x)) + \int_{\tau_\ell}^t \left[\langle \nabla u_{m_\ell}(X_r(x)), b(X_r(x)) \rangle + \tfrac{1}{2}\Delta u_{m_\ell}(X_r(x)) \right] dr \\
& + \int_{\tau_\ell}^t \nabla u_{m_\ell}(X_r(x)) \, dw_r + e^1 \left(l_t(x) - l_{\tau_\ell}(x) \right) \\
=& u_{m_\ell}(X_{\tau_\ell}(x)) + \int_{\tau_\ell}^t b_{m_\ell}(X_r(x)) \, dr + \int_{\tau_\ell}^t \sigma_{m_\ell}(X_r(x)) \, dw_r + e^1 \left(l_t(x) - l_{\tau_\ell}(x) \right).
\end{aligned} \qquad (2.7)$$

For every ℓ we define a continuous semimartingale $(M_t^\ell(x))_t$ by

$$M_t^\ell(x) := \begin{cases} 0 & \text{if } t \in [0, \tau_\ell], \\ \int_{\tau_\ell}^t b_{m_\ell}^1(X_r(x)) \, dr + \int_{\tau_\ell}^t \sigma_{m_\ell}^1(X_r(x)) \, dw_r & \text{if } t \in [\tau_\ell, \tau_{\ell+1}], \\ M_{\tau_{\ell+1}}^\ell(x) & \text{if } t \geq \tau_{\ell+1}. \end{cases} \qquad (2.8)$$

Furthermore, we set $L_t(x) := l_t(x) - l_{\tau_\ell}(x)$ if $t \in [\tau_\ell, \tau_{\ell+1}]$, $\ell \geq 0$, so that

$$u_{m_\ell}^1(X_t(x)) = u_{m_\ell}^1(X_{\tau_\ell}(x)) + M_t^\ell(x) + L_t(x), \qquad t \in [\tau_\ell, \tau_{\ell+1}]. \qquad (2.9)$$

By the Girsanov Theorem there exists a probability measure $\tilde{\mathbb{P}}_\ell(x)$, which is equivalent to $\mathbb{P}$ and under which $M^\ell(x)$ is a continuous martingale. The quadratic variation process is given by

$$[M^\ell(x)]_t = \int_{\tau_\ell}^t \|\sigma_{m_\ell}^1(X_r(x))\|^2 \, dr, \qquad t \in [\tau_\ell, \tau_{\ell+1}],$$

which is strictly increasing in t on $[\tau_\ell, \tau_{\ell+1}]$. We set $\rho_t^\ell := \inf\{s : [M^\ell(x)]_s > t\}$. We can apply the Dambis-Dubins-Schwarz Theorem, in particular its extension in Theorem V.1.7 in [56], since in our case the limit $\lim_{t \to \infty} [M^\ell(x)]_t = [M^\ell(x)]_{\tau_{\ell+1}} < \infty$ is existing, to conclude that the process

$$B_t^\ell(x) := M_{\rho_t^\ell}^\ell(x) \quad \text{for } t < [M^\ell(x)]_{\tau_{\ell+1}}, \qquad B_t^\ell(x) := M_{\tau_{\ell+1}}^\ell(x) \quad \text{for } t \geq [M^\ell(x)]_{\tau_{\ell+1}}, \qquad (2.10)$$

is a $\tilde{\mathbb{P}}_\ell(x)$-Brownian motion w.r.t. the time-changed filtration stopped at time $[M^\ell(x)]_{\tau_{\ell+1}}$ and we have $M_t^\ell(x) = B_{[M^\ell(x)]_t}^\ell$ for all $t \in [\tau_\ell, \tau_{\ell+1}]$. In particular, on $[\tau_\ell, \tau_{\ell+1}]$ the path of $M^\ell(x)$ attains a.s. its minimum at a unique time.

2.2.5 Example: Processes in the Unit Ball

We end this section by considering the example of the unit ball to illustrate our results. Let the domain $G = B_1(0)$ be the closed unit ball in $\mathbb{R}^d$. Then, for $x \in \partial G$, the inner normal field is given by $n(x) = -x$ and the orthogonal projection onto the tangent space by $\pi_x(z) = z - \langle z, x \rangle x$, $z \in \mathbb{R}^d$. The Skorohod equation can be written as

$$X_t(x) = x + \int_0^t b(X_r(x)) \, dr + w_t - \int_0^t X_r(x) \, dl_r(x), \qquad t \geq 0,$$

$$X_t(x) \in G, \quad dl_t(x) \geq 0, \quad \int_0^\infty 1_{\{\|X_t(x)\| < 1\}} \, dl_t(x) = 0, \qquad t \geq 0,$$

and the system describing the derivatives becomes

$$\eta_t = v + \int_0^t Db(X_r(x)) \cdot \eta_r \, dr, \qquad \text{if } t < \inf C,$$

$$\eta_t = \eta_{r(t)-} - \langle \eta_{r(t)-}, X_{r(t)}(x) \rangle X_{r(t)}(x) + \int_{r(t)}^t Db(X_r(x)) \cdot \eta_r \, dr, \qquad \text{if } t \geq \inf C.$$

In this example the nonnegative function $u(x) := \frac{1}{2}(1 - \|x\|^2)$, defined globally on G, can be chosen as the first component u_m^1 of the coordinate mappings u_m for all m. Thus, for the analogue to (2.9) we get

$$u(X_t(x)) = u(x) - \int_0^t \langle X_r(x), b(X_r(x)) \rangle \, dr - \int_0^t X_r(x) \, dw_r - \frac{d}{2}t + l_t(x)$$
$$=: u(x) + M_t(x) + l_t(x).$$

Finally, the quadratic variation process of the semimartingale M^ℓ is given by

$$[M(x)]_t = \int_0^t \|X_r(x)\|^2 \, dr, \qquad t \geq 0.$$

2.3 Proof of the Main Result

2.3.1 Lipschitz Continuity w.r.t. the Initial Datum

The first step in order to prove the differentiability results is to show the Lipschitz continuity of $x \mapsto (X_t(x))_t$ w.r.t. the sup-norm topology on a finite time interval.

Proposition 2.8. *Let T be an arbitrary positive finite stopping time and let $(X_t(x))$ and $(X_t(y))$, $t \geq 0$, be two solutions of (2.1) for some $x, y \in G$. Then, there exists a positive constant c only depending on T such that a.s.*

$$\sup_{t \in [0,T]} \|X_t(x) - X_t(y)\| \leq \|x - y\| \exp(c(T + l_T(x) + l_T(y))).$$

2.3 Proof of the Main Result

Proof. The case $x = y$ is clear and it suffices to consider the case $T < \inf\{t : X_t(x) = X_t(y)\}$. We shall proceed similarly to Lemma 3.8 in [16]. Since ∂G is C^2-smooth, there exists a positive constant $c_1 < \infty$ such that for all $x \in \partial G$ and all $y \in G$,

$$\langle x - y, n(x) \rangle \leq c_1 \|x - y\|^2. \tag{2.11}$$

Let $T_0 := 0$ and for $k \geq 1$,

$$T_k := \inf\left\{ t \geq T_{k-1} : \|X_t(x) - X_t(y)\| \notin \left(\tfrac{1}{2}\|X_{T_{k-1}}(x) - X_{T_{k-1}}(y)\|, 2\|X_{T_{k-1}}(x) - X_{T_{k-1}}(y)\| \right) \right\} \\ \wedge T.$$

Then, by Itô's formula we obtain for any $k \geq 1$ and $t \in (T_{k-1}, T_k]$,

$$\|X_t(x) - X_t(y)\| - \|X_{T_{k-1}}(x) - X_{T_{k-1}}(y)\|$$
$$= \int_{T_{k-1}}^t \frac{\langle X_r(x) - X_r(y), b(X_r(x)) - b(X_r(y)) \rangle}{\|X_r(x) - X_r(y)\|} dr$$
$$+ \int_{T_{k-1}}^t \frac{\langle X_r(x) - X_r(y), n(X_r(x)) \rangle}{\|X_r(x) - X_r(y)\|} dl_r(x) + \int_{T_{k-1}}^t \frac{\langle X_r(y) - X_r(x), n(X_r(y)) \rangle}{\|X_r(x) - X_r(y)\|} dl_r(y)$$
$$\leq c_2 \int_{T_{k-1}}^t \|X_r(x) - X_r(y)\| dr + c_1 \int_{T_{k-1}}^t \|X_r(x) - X_r(y)\| (dl_r(x) + dl_r(y))$$
$$\leq c_3 \|X_{T_{k-1}}(x) - X_{T_{k-1}}(y)\| \int_{T_{k-1}}^{T_k} (dr + dl_r(x) + dl_r(y)),$$

where we have used (2.11) and the Lipschitz continuity of b. Hence, for any $t \in (T_{k-1}, T_k]$,

$$\frac{\|X_t(x) - X_t(y)\|}{\|X_{T_{k-1}}(x) - X_{T_{k-1}}(y)\|} \leq 1 + c_3 \left(T_k - T_{k-1} + l_{T_k}(x) - l_{T_{k-1}}(x) + l_{T_k}(y) - l_{T_{k-1}}(y) \right)$$
$$\leq \exp\left(c_3 \left(T_k - T_{k-1} + l_{T_k}(x) - l_{T_{k-1}}(x) + l_{T_k}(y) - l_{T_{k-1}}(y) \right) \right),$$

and

$$\frac{\|X_t(x) - X_t(y)\|}{\|x - y\|} = \frac{\|X_t(x) - X_t(y)\|}{\|X_{T_{k-1}}(x) - X_{T_{k-1}}(y)\|} \prod_{j=1}^{k-1} \frac{\|X_{T_j}(x) - X_{T_j}(y)\|}{\|X_{T_{j-1}}(x) - X_{T_{j-1}}(y)\|}$$
$$\leq \prod_{j=1}^k \exp\left(c_3 \left(T_j - T_{j-1} + l_{T_j}(x) - l_{T_{j-1}}(x) + l_{T_j}(y) - l_{T_{j-1}}(y) \right) \right)$$
$$\leq \exp\left(c_3 \left(T_k + l_{T_k}(x) + l_{T_k}(y) \right) \right)$$
$$\leq \exp\left(c_3 \left(T + l_T(x) + l_T(y) \right) \right),$$

which proves the proposition. □

Remark 2.9. For an arbitrary positive finite stopping time T and every $\ell \geq 0$ we define the stopping time

$$\hat{\tau}_\ell(y) := \inf\{t > \tau_\ell : \|X_t(x) - X_t(y)\| \geq \tfrac{\delta_0}{2}\} \wedge \tau_{\ell+1} \wedge T, \qquad y \in G,$$

with δ_0 as in Section 2.2.4. Then, by the definition of τ_ℓ we have that $X_t(y) \in U_{m_l}$ for all $t \in [\tau_\ell, \hat{\tau}_\ell(y)]$. Thus, the process $(M_t^\ell(y))_{0 \le t \le T}$, $y \in G$, can be defined similarly to $M^\ell(x)$ by

$$M_t^\ell(y) := \begin{cases} 0 & \text{if } t \in [0, \tau_\ell], \\ \int_{\tau_\ell}^t b_{m_\ell}^1(X_r(y))\, dr + \int_{\tau_\ell}^t \sigma_{m_\ell}^1(X_r(y))\, dw_r & \text{if } t \in [\tau_\ell, \hat{\tau}_\ell(y)], \\ M_{\hat{\tau}_\ell(y)}^\ell(y) & \text{if } t \ge \hat{\tau}_\ell(y). \end{cases}$$

By Proposition 2.8 there exists for every finite stopping time $T > 0$ a random $\Delta_T > 0$ such that

$$\sup_{t \in [0,T]} \|X_t(x) - X_t(y)\| < \frac{\delta_0}{2}, \qquad \forall y \in B_{\Delta_T}(x) \cap G,$$

i.e. for such y we have $\hat{\tau}_\ell(y) = \tau_{\ell+1} \wedge T$.

In the next lemma we collect some immediate consequences of Proposition 2.8.

Lemma 2.10. *Fix some arbitrary real $T > 0$ and let Δ_T and $\hat{\tau}_\ell(y)$ be as in Remark 2.9. Then, for every $\ell \ge 0$ we have:*

i) For all $x, y \in G$ and $p > 1$,

$$\mathbb{E}\left[\sup_{t \in [0,T]} \left|M_t^\ell(x) - M_t^\ell(y)\right|^p \mathbb{1}_{\{\|x-y\| < \Delta_T\}}\right] \le c \|x - y\|^p$$

for some positive constant $c = c(p, T)$. There exists a modification of $x \mapsto (M_t^\ell(x))_{t \in [0,T]}$ which is continuous w.r.t. the sup-norm topology.

ii) For all $x \in G$ and $s, t \in [0, T]$ and $p > 1$,

$$\mathbb{E}\left[\left|M_t^\ell(x) - M_s^\ell(x)\right|^p\right] \le c |t - s|^{p/2},$$

for some positive constant $c = c(p, T)$ not depending on x. There exists a modification of $(M_t^\ell(x))_{t \in [0,T]}$ which is Hölder continuous of order α for every $\alpha \in (0, \frac{1}{2})$ such that

$$\sup_{x \in G} \mathbb{E}\left[\sup_{s \ne t} \left(\frac{|M_t^\ell(x) - M_s^\ell(x)|}{|t - s|^\alpha}\right)^p\right] < \infty.$$

iii) We set for abbreviation $\tilde{M}_t^\ell(x, y) := (M_{t \wedge \hat{\tau}_\ell(y)}^\ell(y) - M_{t \wedge \hat{\tau}_\ell(y)}^\ell(x))$, $t \in [0, T]$, $x, y \in G$. Let $x \in G$ and $(x_i)_{i \in I} \subset G$ be a family of points in G such that $x_i \ne x$ for all $i \in I$. Then, for all $s, t \in [0, T]$ and $p > 1$ we have

$$\sup_{i \in I} \frac{1}{\|x_i - x\|^p} \mathbb{E}\left[\left|\tilde{M}_t^\ell(x, x_i) - \tilde{M}_s^\ell(x, x_i)\right|^p\right] \le c |t - s|^{p/2},$$

for some positive constant $c = c(p, T)$. Moreover, for every $i \in I$ there exists a modification of $(\tilde{M}_t^\ell(x, x_i))_{t \in [0,T]}$ which is Hölder continuous of order α for every $\alpha \in (0, \frac{1}{2})$ such that

$$\mathbb{E}\left[\sup_{s \ne t} \left(\frac{|\tilde{M}_t^\ell(x, x_i) - \tilde{M}_s^\ell(x, x_i)|}{|t - s|^\alpha}\right)^p\right] \le c \|x_i - x\|^p$$

with some positive constant c not depending on x_i.

2.3 Proof of the Main Result

Proof. i) It follows directly from Proposition 2.8, the uniform Lipschitz continuity of b_m^1 and σ_m^1 and the Burkholder inequality that

$$\mathbb{E}\left[\sup_{t\in[0,\hat{\tau}_\ell(y)]}\left|M_t^\ell(x) - M_t^\ell(y)\right|^p\right] \leq c_1 \left\|x - y\right\|^p$$

for some positive constant c_1. Note that the random constant $\exp(c(T+l_T(x)+l_T(y)))$, appearing in Proposition 2.8, has finite expectation. Since by the definition of Δ_T,

$$\sup_{t\in[0,T]}\left|M_t^\ell(x) - M_t^\ell(y)\right|^p \mathbb{1}_{\{\|x-y\|<\Delta_T\}} \leq \sup_{t\in[0,\hat{\tau}_\ell(y)]}\left|M_t^\ell(x) - M_t^\ell(y)\right|^p,$$

this implies the estimate in i). To prove now the existence of a continuous modification one only needs to modify slightly the proof of Kolmogorov's continuity theorem (cf. e.g. Theorem I.2.1 in [56]).

ii) The estimate is clear again by Burkholder's inequality. Note that the functions b_m^1 and σ_m^1 are uniformly bounded so that the constant does not depend on x. Furthermore, again by the proof of Kolmogorov's continuity theorem it follows that in this case the L^p norm of the Hölder norm of the modification does also not depend on x.

iii) In the following the symbol c denotes a constant whose value may change from one occurence to the other one. Let $0 \leq s \leq t \leq T$ and set $\hat{s}_i := s \wedge \hat{\tau}_\ell(x_i)$ and $\hat{t}_i := t \wedge \hat{\tau}_\ell(x_i)$. Then, $|\hat{t}_i - \hat{s}_i| \leq |t - s|$ for all i. By the definition of M^ℓ and $\tilde{M}^\ell$ we have

$$\sup_{i\in I}\frac{1}{\|x_i - x\|^p}\mathbb{E}\left[\left|\tilde{M}_t^\ell(x, x_i) - \tilde{M}_s^\ell(x, x_i)\right|^p\right]$$

$$\leq c \sup_{i\in I}\frac{1}{\|x_i - x\|^p}\mathbb{E}\left[\left|\int_{\hat{s}_i}^{\hat{t}_i}(b_{m_\ell}^1(X_r(x_i)) - b_{m_\ell}^1(X_r(x))))\, dr\right|^p\right]$$

$$+ c \sup_{i\in I}\frac{1}{\|x_i - x\|^p}\mathbb{E}\left[\left|\int_{\hat{s}_i}^{\hat{t}_i}(\sigma_{m_\ell}^1(X_r(x_i)) - \sigma_{m_\ell}^1(X_r(x)))\, dw_r\right|^p\right].$$

Using the uniform Lipschitz continuity of b_m and Proposition 2.8 the first term can be estimated by

$$c \sup_{i\in I}\left(\frac{|t - s|}{\|x_i - x\|}\right)^p \mathbb{E}\left[\sup_{r\in[\hat{s}_i, \hat{t}_i]}\|X_r(x_i) - X_r(x)\|^p\right] \leq c |t - s|^p.$$

For the second term we get the following estimate by Burkholder's inequality, the uniform Lipschitz continuity of σ_m and again by Proposition 2.8:

$$c \sup_{i\in I}\frac{1}{\|x_i - x\|^p}\mathbb{E}\left[\sup_{r\in[\hat{s}_i, \hat{t}_i]}\left|\int_{\hat{s}_i}^{r}\left(\sigma_{m_\ell}^1(X_r(x_i)) - \sigma_{m_\ell}^1(X_r(x))\right)\, dw_r\right|^p\right]$$

$$\leq c \sup_{i\in I}\frac{1}{\|x_i - x\|^p}\mathbb{E}\left[\left(\int_{\hat{s}_i}^{\hat{t}_i}\|X_r(x_i) - X_r(x)\|^2 dr\right)^{p/2}\right]$$

$$\leq c \sup_{i\in I}\frac{|t - s|^{p/2}}{\|x_i - x\|^p}\mathbb{E}\left[\sup_{r\in[\hat{s}_i, \hat{t}_i]}\|X_r(x_i) - X_r(x)\|^p\right]$$

$$\leq c |t - s|^{p/2}$$

and we obtain the desired estimate. The existence of a Hölder continuous modification and the uniform L^p-bound of its Hölder norm follow again from the Kolmogorov criterion.

□

2.3.2 Convergence of the Local Time Processes

We fix from now on an arbitrary $T > 0$. In the following let (A_n) be the family of connected components of $[0,T]\backslash C$ and q_n be as above. Furthermore, let $(\tau_\ell)_\ell$ be the sequence of stopping times defined as before by

$$\tau_0 := 0, \quad \tau_{\ell+1} := \{t > \tau_\ell : \text{dist}(X_t(x), \partial U_{m_\ell}) < \delta_0\} \wedge T, \quad \ell \geq 0.$$

We may suppose that the q_n are chosen in such a way that for every n we have $[r(q_n), q_n] \subset [\tau_\ell, \tau_{\ell+1}]$ if $r(q_n) \in [\tau_\ell, \tau_{\ell+1}]$ for some ℓ.

In order to compute the local time $l(x)$, recall that on every interval $[\tau_\ell, \tau_{\ell+1}]$, $\ell \geq 0$, $l(x)$ is carried by the set of times t, when $u^1_{m_\ell}(X_t(x)) = 0$. Therefore, we can apply Skorohod's Lemma (see e.g. Lemma VI.2.1 in [56]) to equation (2.9) to obtain

$$L_t(x) = \left[-u^1_{m_\ell}(X_{\tau_\ell}(x)) - \inf_{\tau_\ell \leq s \leq t} M^\ell_s(x) \right]^+, \quad t \in [\tau_\ell, \tau_{\ell+1}].$$

Fix any $q_n > \inf C$ and ℓ such that $q_n \in [\tau_\ell, \tau_{\ell+1}]$. Since $u^1_{m_\ell}(X_{r(q_n)}(x)) = 0$ and $t \mapsto L_t(x)$ is non-decreasing, we have for all $\tau_\ell \leq s \leq r(q_n)$:

$$M^\ell_{r(q_n)}(x) = -u^1_{m_\ell}(X_{\tau_\ell}(x)) - L_{r(q_n)}(x) \leq -u^1_{m_\ell}(X_{\tau_\ell}(x)) - L_s(x)$$
$$= -u^1_{m_\ell}(X_s(x)) + M^\ell_s(x) \leq M^\ell_s(x).$$

Moreover, $L(x)$ is constant on $[r(q_n), t]$ for all $t \in A_n \cap [\tau_\ell, \tau_{\ell+1}]$, so that

$$L_t(x) = L_{r(q_n)}(x) = \left[-u^1_{m_\ell}(X_{\tau_\ell}(x)) - M^\ell_{r(q_n)}(x) \right]^+, \quad t \in A_n \cap [\tau_\ell, \tau_{\ell+1}]. \tag{2.12}$$

Note that $r(q_n)$ is the unique time in $[\tau_\ell, q_n]$, when $M^\ell(x)$ attains its minimum. Analogously we compute the local time of the process with perturbed starting point. For fixed $v \in \mathbb{R}^d$ we set $x_\varepsilon := x + \varepsilon v$, $\varepsilon \in \mathbb{R}$, where $|\varepsilon|$ is always supposed to be sufficiently small, such that x_ε lies in G. Furthermore, there exists a random $\Delta_n > 0$ such that $X_{\tau_\ell}(x_\varepsilon) \in U_{m_\ell}$ and $\hat{\tau}_\ell(x_\varepsilon) > q_n$ for all $\varepsilon \in (-\Delta_n, \Delta_n)$ (cf. Remark 2.9). As above we obtain for such ε:

$$L_{q_n}(x_\varepsilon) = L_{r_\varepsilon(q_n)}(x_\varepsilon) = \left[-u^1_{m_\ell}(X_{\tau_\ell}(x_\varepsilon)) - M^\ell_{r_\varepsilon(q_n)}(x_\varepsilon) \right]^+, \tag{2.13}$$

where $r_\varepsilon(q_n)$, defined similarly as $r(q_n)$, is the unique time in $[\tau_\ell, q_n]$, when $M^\ell(x_\varepsilon)$ attains its minimum.

Lemma 2.11. *For all q_n we have $r_\varepsilon(q_n) \to r(q_n)$ a.s. for $\varepsilon \to 0$.*

2.3 Proof of the Main Result

Proof. We fix some q_n and ℓ such that $q_n \in [\tau_\ell, \tau_{\ell+1}]$. For every sequence $(\varepsilon_k)_k$ converging to zero we can extract a subsequence of $(r_{\varepsilon_k}(q_n))$, still denoted by $(r_{\varepsilon_k}(q_n))$, converging to some $\hat{r}(q_n)$. By construction we have a.s.

$$M^\ell_{r_{\varepsilon_k}(q_n)}(x_{\varepsilon_k}) \leq M^\ell_{r(q_n)}(x_{\varepsilon_k})$$

for every k. Note that on one hand the right hand side converges to $M^\ell_{r(q_n)}(x)$ as $k \to \infty$ by Lemma 2.10 i). On the other hand the left hand side converges to $M^\ell_{\hat{r}(q_n)}(x)$, since

$$\left| M^\ell_{r_{\varepsilon_k}(q_n)}(x_{\varepsilon_k}) - M^\ell_{\hat{r}(q_n)}(x) \right| \leq \left| M^\ell_{r_{\varepsilon_k}(q_n)}(x_{\varepsilon_k}) - M^\ell_{r_{\varepsilon_k}(q_n)}(x) \right| + \left| M^\ell_{r_{\varepsilon_k}(q_n)}(x) - M^\ell_{\hat{r}(q_n)}(x) \right|$$

$$\leq \sup_{t \in [0,T]} \left| M^\ell_t(x_{\varepsilon_k}) - M^\ell_t(x) \right| + \left| M^\ell_{r_{\varepsilon_k}(q_n)}(x) - M^\ell_{\hat{r}(q_n)}(x) \right|,$$

which tends to zero for $k \to \infty$ by Lemma 2.10 i) and ii). Thus, $M^\ell_{\hat{r}(q_n)}(x) \leq M^\ell_{r(q_n)}(x)$. Since $r(q_n)$ is unique time in $[\tau_\ell, q_n]$, when $M^\ell(x)$ attains its minimum, this implies $\hat{r}(q_n) = r(q_n)$. $\square$

Lemma 2.12. *Let $(W_t)_{t \geq 0}$ be a Brownian motion on $(\Omega, \mathcal{F}, \mathbb{P})$. For all $T > 0$, let $\vartheta : \Omega \to [0, T]$ be the random variable such that a.s.*

$$W_\vartheta < W_s, \quad \forall s \in [0, T] \setminus \{\vartheta\}.$$

Then,

$$\liminf_{s \to \vartheta} \frac{W_s - W_\vartheta}{\sqrt{|s - \vartheta|} h(|s - \vartheta|)} \geq 1 \quad a.s.,$$

for every function h on $[0, \infty)$ satisfying $0 < h(t) \downarrow 0$ as $t \downarrow 0$ and $\int_0^{r_0} h(t) \frac{dt}{t} < \infty$ for some $r_0 > 0$.

Proof. It suffices to consider the case $T = 1$. We recall the following path decomposition of a Brownian motion, proven in [25]. Denoting by $(M, \hat{M})$ two independent copies of the standard Brownian meander (see [56]), we set for all $r \in (0, 1)$,

$$V_r(t) := \begin{cases} -\sqrt{r} M(1) + \sqrt{r} M(\frac{r-t}{r}), & t \in [0, r] \\ -\sqrt{r} M(1) + \sqrt{1-r} \hat{M}(\frac{t-r}{1-r}), & t \in (r, 1] \end{cases}$$

Let now $(\tau, M, \hat{M})$ be an independent triple, such that τ has the arcsine law. Then, $V_\tau \stackrel{d}{=} W$. This formula has the following meaning: τ is the unique time in $[0, 1]$, when the path attains minimum $-\sqrt{\tau} M(1)$. The path starts in zero at time $t = 0$ and runs backward the path of M on $[0, \tau]$ and then it runs the path of $\hat{M}$. Moreover, it was proved in [43] that the law of the Brownian meander is absolutely continuous w.r.t. the law of the three-dimensional Bessel process $(R_t)_{t \geq 0}$ on the time interval $[0, 1]$ starting in zero. We recall that a.s.

$$\liminf_{t \to 0} \frac{R_t}{\sqrt{t} h(t)} \geq 1$$

for every function h satisfying the conditions in the statement (see [44], p. 164). Since the same asymptotics hold for the Brownian meander at zero, the claim follows.
$\square$

Proposition 2.13. *Let $\delta > 0$ be arbitrary. Then, for all q_n there exists a random $\Delta_n > 0$ such that*
$$\mathbb{E}\left[\left(\frac{|r_\varepsilon(q_n) - r(q_n)|^\delta}{|\varepsilon|}\right)^p \mathbb{1}_{\{0 < |\varepsilon| < \Delta_n\}}\right] \leq c < \infty \tag{2.14}$$
for every $p > 1$ and a constant $c = c(\delta, p)$ not depending on ε. In particular, we have for every $\delta > 0$ and $p > 1$
$$\frac{|r_\varepsilon(q_n) - r(q_n)|^\delta}{\varepsilon} \longrightarrow 0 \quad \text{in } L^p \text{ as } \varepsilon \to 0.$$

Proof. To prove (2.14) it is enough to consider the case $\delta < 1$. By construction we have for every q_n and ℓ such that $q_n \in [\tau_\ell, \tau_{\ell+1}]$,
$$M^\ell_{r_\varepsilon(q_n)}(x_\varepsilon) \leq M^\ell_{r(q_n)}(x_\varepsilon).$$
Since for ε small enough $q_n < \hat{\tau}_\ell(x_\varepsilon)$ (see Remark 2.9) we have $M^\ell_t(x_\varepsilon) = M^\ell_t(x) + \tilde{M}^\ell_t(x, x_\varepsilon)$ for every $t \in [\tau_\ell, q_n]$ with $\tilde{M}^\ell(x, x_\varepsilon)$ defined as in Lemma 2.10 iii), this is equivalent to
$$M^\ell_{r_\varepsilon(q_n)}(x) - M^\ell_{r(q_n)}(x) \leq \tilde{M}^\ell_{r(q_n)}(x, x_\varepsilon) - \tilde{M}^\ell_{r_\varepsilon(q_n)}(x, x_\varepsilon),$$
which implies
$$\frac{M^\ell_{r_\varepsilon(q_n)}(x) - M^\ell_{r(q_n)}(x)}{|r_\varepsilon(q_n) - r(q_n)|^{(1-\delta)/2}} \mathbb{1}_{\{r_\varepsilon(q_n) \neq r(q_n)\}} \leq \frac{\left|\tilde{M}^\ell_{r(q_n)}(x, x_\varepsilon) - \tilde{M}^\ell_{r_\varepsilon(q_n)}(x, x_\varepsilon)\right|}{|r_\varepsilon(q_n) - r(q_n)|^{(1-\delta)/2}} \mathbb{1}_{\{r_\varepsilon(q_n) \neq r(q_n)\}}. \tag{2.15}$$
Recall that $M^\ell_\cdot(x) = B^\ell_{[M^\ell(x)]_\cdot}$, where B^ℓ is a $\tilde{\mathbb{P}}_\ell(x)$-Brownian motion (see (2.10)) and B^ℓ attains its minimum over $\left[[M^\ell(x)]_{\tau_\ell}, [M^\ell(x)]_{q_n}\right]$ at time $[M^\ell(x)]_{r(q_n)}$. Hence, applying Lemma 2.12 with $h(t) = t^{\delta/2}$ it follows that
$$M^\ell_{r_\varepsilon(q_n)}(x) - M^\ell_{r(q_n)}(x) = B^\ell_{[M^\ell(x)]_{r_\varepsilon(q_n)}} - B^\ell_{[M^\ell(x)]_{r(q_n)}} \geq \tfrac{1}{2}\left|[M^\ell(x)]_{r_\varepsilon(q_n)} - [M^\ell(x)]_{r(q_n)}\right|^{(1+\delta)/2}$$
$$= \tfrac{1}{2}\left|\int_{r(q_n)}^{r_\varepsilon(q_n)} \|\sigma^1_{m_\ell}(X_r(x))\|^2 \, dr\right|^{(1+\delta)/2}$$
for all $\varepsilon \in (-\Delta_n, \Delta_n)$ for some positive Δ_n. Since
$$\|\sigma^1_{m_\ell}(X_{r(q_n)}(x))\|^2 = \|\nabla u^1_{m_\ell}(X_{r(q_n)}(x))\|^2 = \|n(X_{r(q_n)}(x))\|^2 = 1,$$
we have by Lemma 2.11, possibly after choosing a smaller Δ_n, that $\|\sigma^1_{m_\ell}(X_r(x))\|^2$ is bounded away from zero uniformly in r between $r(q_n)$ and $r_\varepsilon(q_n)$. Thus,
$$M^\ell_{r_\varepsilon(q_n)}(x) - M^\ell_{r(q_n)}(x) \geq c_1 \, |r_\varepsilon(q_n) - r(q_n)|^{(1+\delta)/2}$$
and we derive from (2.15) that
$$c_1 \, |r_\varepsilon(q_n) - r(q_n)|^\delta \leq \sup_{\substack{s,t \in [0,T] \\ s \neq t}} \frac{\left|\tilde{M}^\ell_t(x, x_\varepsilon) - \tilde{M}^\ell_s(x, x_\varepsilon)\right|}{|t - s|^{(1-\delta)/2}}.$$

Finally, we get for every $p > 1$ using Lemma 2.10 iii)

$$\mathbb{E}\left[|r_\varepsilon(q_n) - r(q_n)|^{\delta p} \, \mathbb{1}_{\{0<|\varepsilon|<\Delta_n\}}\right] \leq c_2 \, \mathbb{E}\left[\sup_{s\neq t}\left(\frac{|\tilde{M}_t^\ell(x,x_\varepsilon) - \tilde{M}_s^\ell(x,x_\varepsilon)|}{|t-s|^{(1-\delta)/2}}\right)^p\right] \leq c_3 \, |\varepsilon|^p,$$

and we obtain (2.14). Since δ is arbitrary, the L^p-convergence follows immediately from (2.14) by Hölder's inequality and Lemma 2.11. □

Corollary 2.14. *For every q_n and ℓ such that $q_n \in [\tau_\ell, \tau_{\ell+1}]$ we have*

i) $\frac{1}{\varepsilon}\left|M_{r(q_n)}^\ell(x_\varepsilon) - M_{r_\varepsilon(q_n)}^\ell(x_\varepsilon)\right| \longrightarrow 0$ a.s. as $\varepsilon \to 0$,

ii) $\frac{1}{\varepsilon}\left|l_{r(q_n)}(x_\varepsilon) - l_{r_\varepsilon(q_n)}(x_\varepsilon)\right| \longrightarrow 0$ a.s. as $\varepsilon \to 0$.

Proof. ii) follows from i). Indeed, by Proposition 2.8 and Lemma 2.11 we can choose ε so small that $l_{r(q_n)}(x_\varepsilon) = l_{r_\varepsilon(q_n)}(x_\varepsilon) = 0$ if $q_n < \inf C$ and $l_{r(q_n)}(x_\varepsilon), l_{r_\varepsilon(q_n)}(x_\varepsilon) > 0$ if $q_n > \inf C$. In the first case ii) is trivial and the latter case we have by (2.13)

$$\left|l_{r(q_n)}(x_\varepsilon) - l_{r_\varepsilon(q_n)}(x_\varepsilon)\right| = \left|L_{r(q_n)}(x_\varepsilon) - L_{r_\varepsilon(q_n)}(x_\varepsilon)\right| = M_{r_\varepsilon(r(q_n))}^\ell(x_\varepsilon) - M_{r_\varepsilon(q_n)}^\ell(x_\varepsilon)$$
$$\leq M_{r(q_n)}^\ell(x_\varepsilon) - M_{r_\varepsilon(q_n)}^\ell(x_\varepsilon), \qquad (2.16)$$

where we have used the fact that $M^\ell(x_\varepsilon)$ attains its minimum over $[\tau_\ell, q_n]$ at time $r_\varepsilon(q_n)$ and its minimum over $[\tau_\ell, r(q_n)]$ at time $r_\varepsilon(r(q_n))$, respectively. Therefore it suffices to prove i). In a first step we show that for an arbitrary $\delta > 0$ and for every $p > 1$

$$\frac{\left|M_{r(q_n)}^\ell(x_\varepsilon) - M_{r_\varepsilon(q_n)}^\ell(x_\varepsilon)\right|^\delta}{\varepsilon} \longrightarrow 0 \quad \text{in } L^p. \qquad (2.17)$$

It is enough to consider p such that $\delta p > 1$. Then, for any $\alpha \in (0, \frac{1}{2})$,

$$\mathbb{E}\left[\frac{1}{\varepsilon^p}\left|M_{r(q_n)}^\ell(x_\varepsilon) - M_{r_\varepsilon(q_n)}^\ell(x_\varepsilon)\right|^{\delta p}\right]$$
$$=\mathbb{E}\left[\frac{|r_\varepsilon(q_n) - r(q_n)|^{\alpha \delta p}}{\varepsilon^p}\left(\frac{\left|M_{r(q_n)}^\ell(x_\varepsilon) - M_{r_\varepsilon(q_n)}^\ell(x_\varepsilon)\right|}{|r_\varepsilon(q_n) - r(q_n)|^\alpha}\right)^{\delta p} \mathbb{1}_{\{r_\varepsilon(q_n)\neq r(q_n)\}}\right]$$
$$\leq \mathbb{E}\left[\frac{|r_\varepsilon(q_n) - r(q_n)|^{2\alpha\delta p}}{\varepsilon^{2p}}\right]^{1/2} \mathbb{E}\left[\sup_{s\neq t}\left(\frac{|M_t^\ell(x_\varepsilon) - M_s^\ell(x_\varepsilon)|}{|t-s|^\alpha}\right)^{2\delta p}\right]^{1/2}.$$

The first term tends to zero by Proposition 2.13, while the second term is uniformly bounded by Lemma 2.10 ii), and we obtain (2.17). We prove now i). Assume $\frac{1}{\varepsilon}\left|M_{r(q_n)}^\ell(x_\varepsilon) - M_{r_\varepsilon(q_n)}^\ell(x_\varepsilon)\right| \not\to 0$ with positive probability. Since $r_\varepsilon(q_n) \to r(q_n)$ a.s. by Lemma 2.11 and M^ℓ is continuous in t and x, it follows that a.s. $\left|M_{r(q_n)}^\ell(x_\varepsilon) - M_{r_\varepsilon(q_n)}^\ell(x_\varepsilon)\right| \to 0$. Thus, $\frac{1}{\varepsilon}\left|M_{r(q_n)}^\ell(x_\varepsilon) - M_{r_\varepsilon(q_n)}^\ell(x_\varepsilon)\right|^\delta \to \infty$ with positive probability for any $\delta < 1$, which contradicts (2.17). □

2.3.3 Proof of the Differentiability

Conververgence along subsequences

We shall proceed similarly to Step 5 in the proof of Theorem 1 in [26]. Denote by $\eta_t(\varepsilon) := \frac{1}{\varepsilon}(X_t(x_\varepsilon) - X_t(x))$ the difference quotient, $x_\varepsilon = x + \varepsilon v$ for any fixed $v \in \mathbb{R}^d$. Let now $t \in [0,T]\backslash C$ and let n be such that $t \in A_n$. Using Proposition 2.8 there exists $\Delta_n > 0$ such that a.s. $l_{q_n}(x) = l_{q_n}(x_\varepsilon) = 0$ if $q_n < \inf C$ and both of them are strictly positive if $q_n > \inf C$ for all $|\varepsilon| \in (0, \Delta_n)$. Then,

$$\eta_t(\varepsilon) = \eta_{r(q_n)}(\varepsilon) + \frac{1}{\varepsilon}\int_{r(q_n)}^t (b(X_r(x_\varepsilon)) - b(X_r(x)))\,dr + \frac{1}{\varepsilon}\int_0^{r_\varepsilon(q_n)} n(X_r(x_\varepsilon))\,dl_r(x_\varepsilon)$$

$$- \frac{1}{\varepsilon}\int_0^{r(q_n)} n(X_r(x_\varepsilon))\,dl_r(x_\varepsilon)$$

$$= \eta_{r(q_n)}(\varepsilon) + \sum_{k=1}^d \int_{r(q_n)}^t \left[\int_0^1 \frac{\partial b}{\partial x^k}(X_r^{\alpha,\varepsilon})\,d\alpha\right]\eta_r^k(\varepsilon)\,dr + R_{q_n}(x_\varepsilon), \qquad (2.18)$$

where $X_r^{\alpha,\varepsilon} := \alpha X_r(x_\varepsilon) + (1-\alpha)X_r(x)$, $\alpha \in [0,1]$, and

$$R_{q_n}(x_\varepsilon) := \frac{1}{\varepsilon}\int_{r(q_n)}^{r_\varepsilon(q_n)} n(X_r(x_\varepsilon))\,dl_r(x_\varepsilon). \qquad (2.19)$$

Note that if $q_n < \inf C$ we have $r(q_n) = 0$, $\eta_{r(q_n)}(\varepsilon) = v$ and $R_{q_n}(x_\varepsilon) = 0$. In any case,

$$\|R_{q_n}(x_\varepsilon)\| \leq \frac{1}{\varepsilon}\left|\int_{r(q_n)}^{r_\varepsilon(q_n)} \|n(X_r(x_\varepsilon))\|\,dl_s(x_\varepsilon)\right| = \left|\frac{l_{r_\varepsilon(q_n)}(x_\varepsilon) - l_{r(q_n)}(x_\varepsilon)}{\varepsilon}\right| \longrightarrow 0 \quad \text{as } \varepsilon \to 0, \quad (2.20)$$

by Corollary 2.14. Recall that $\|\eta_t(\varepsilon)\| \leq \exp(c_1(T + l_T(x) + l_T(y)))$ for all $t \in [0,T]$ and $\varepsilon \neq 0$ by Proposition 2.8. Let $(\varepsilon_\nu)_\nu$ be any sequence converging to zero. By a diagonal procedure, we can extract a subsequence $(\nu_l)_l$ such that $\eta_{r(q_n)}(\varepsilon_{\nu_l})$ has a limit $\hat{\eta}_{r(q_n)} \in \mathbb{R}^d$ and $\eta_{\tau_\ell}(\varepsilon_{\nu_l})$ has a limit $\hat{\eta}_{\tau_\ell} \in \mathbb{R}^d$ as $l \to \infty$ for all $n \in \mathbb{N}$ and for all $\ell \geq 0$.

Let now $\hat{\eta} : [0,T]\backslash C \to \mathbb{R}^d$ be the unique solution of

$$\hat{\eta}_t := \hat{\eta}_{r(q_n)} + \int_{r(q_n)}^t Db(X_r(x))\cdot\hat{\eta}_r\,dr, \qquad t \in A_n.$$

By (2.18) and Proposition 2.8, we get for $|\varepsilon| \in (0, \Delta_n)$ and $t \in A_n$,

$$\|\eta_t(\varepsilon) - \hat{\eta}_t\| \leq \|\eta_{r(q_n)}(\varepsilon) - \hat{\eta}_{r(q_n)}\| + \|R_{q_n}(x_\varepsilon)\|$$

$$+ \sup_{r \in A_n}\|Db(X_r^{\alpha,\varepsilon}) - Db(X_r(x))\|\exp(c_1(T + l_T(x) + l_T(y))$$

$$+ c_2\int_0^t \|\eta_r(\varepsilon) - \hat{\eta}_r\|\,dr.$$

Since $\eta_{r(q_n)}(\varepsilon_{\nu_l}) \to \hat{\eta}_{r(q_n)}$, $\|R_{q_n}(x_\varepsilon)\| \to 0$, $X_r^{\alpha,\varepsilon_{\nu_l}} \to X_r(x)$ uniformly in $r \in [0,t]$ and since the derivatives of b are continuous, we obtain by Gronwall's Lemma that $\eta_t(\varepsilon_{\nu_l})$ converges to $\hat{\eta}_t$

2.3 Proof of the Main Result

uniformly in $t \in A_n$ for every n. Thus, since C has zero Lebesgue measure, $\eta_t(\varepsilon_{\nu_l})$ converges to $\hat{\eta}_t$ for all $t \in [0,T]\backslash C$ as $l \to \infty$ and by the dominated convergence theorem we get

$$\hat{\eta}_t = v + \int_0^t Db(X_r(x)) \cdot \hat{\eta}_r^k \, dr, \qquad t \in [0, \inf C),$$

$$\hat{\eta}_t = \hat{\eta}_{r(t)} + \int_{r(t)}^t Db(X_r(x)) \cdot \hat{\eta}_r^k \, dr, \qquad t \in [\inf C, T]\backslash C.$$

Lemma 2.15. *For every $q_n > \inf C$,*

i) $\langle \eta_{r(q_n)}(\varepsilon), n(X_{r(q_n)}(x)) \rangle \to 0$ *a.s. and in L^p, $p > 1$, as $\varepsilon \to 0$,*

ii) $\langle \eta_{r_\varepsilon(q_n)}(\varepsilon), n(X_{r_\varepsilon(q_n)}(x_\varepsilon)) \rangle \to 0$ *a.s. and in L^p, $p > 1$, as $\varepsilon \to 0$.*

Proof. By dominated convergence it suffices to prove convergence almost surely. Let ℓ be such that $q_n \in [\tau_\ell, \tau_{\ell+1}]$. Then, clearly $X_{r(q_n)}(x) \in U_{m_\ell} \cap \partial G$. Recall that $n(X_{r(q_n)}(x)) = \nabla u^1_{m_\ell}(X_{r(q_n)}(x))$, and by Taylor's formula we get

$$\langle \eta_{r(q_n)}(\varepsilon), n(X_{r(q_n)}(x)) \rangle = \frac{1}{\varepsilon} \left(u^1_{m_\ell}(X_{r(q_n)}(x_\varepsilon)) - u^1_{m_\ell}(X_{r(q_n)}(x)) \right) + O(\varepsilon).$$

Note that the term of second order in the Taylor expansion is in $O(\varepsilon)$ by Proposition 2.8. Recall that $u^1_{m_\ell}(X_{r(q_n)}(x)) = 0$, and combining formula (2.9) and (2.13), we get

$$u^1_{m_\ell}(X_{r(q_n)}(x_\varepsilon)) = u^1_{m_\ell}(X_{\tau_\ell}(x_\varepsilon)) + M^\ell_{r(q_n)}(x_\varepsilon) + L_{r(q_n)}(x_\varepsilon) = M^\ell_{r(q_n)}(x_\varepsilon) - M^\ell_{r_\varepsilon(r(q_n))}(x_\varepsilon)$$

for all $\varepsilon \in (-\Delta_n, \Delta_n)$ for some positive Δ_n. Arguing similarly as in (2.16) we obtain from Corollary 2.14 i) that

$$\left| \frac{u^1_{m_\ell}(X_{r(q_n)}(x_\varepsilon)) - u^1_{m_\ell}(X_{r(q_n)}(x))}{\varepsilon} \right| \leq 2 \left| \frac{M^\ell_{r(q_n)}(x_\varepsilon) - M^\ell_{r_\varepsilon(q_n)}(x_\varepsilon)}{\varepsilon} \right| \longrightarrow 0 \qquad \text{as } \varepsilon \to 0,$$

and i) follows. The proof of ii) is rather analogous. For an appropriate $\Delta_n > 0$ we have $r_\varepsilon(q_n) \in [\tau_\ell, \tau_{\ell+1}]$ and $l_{r_\varepsilon(q_n)}(x) > 0$ for all $|\varepsilon| \in (0, \Delta_n)$. Then, for such ε we get again by using Taylor's formula and the fact that $u^1_{m_\ell}(X_{r_\varepsilon(q_n)}(x_\varepsilon)) = 0$,

$$\langle \eta_{r_\varepsilon(q_n)}(\varepsilon), n(X_{r_\varepsilon(q_n)}(x_\varepsilon)) \rangle = \langle \eta_{r_\varepsilon(q_n)}(\varepsilon), \nabla u^1_{m_\ell}(X_{r_\varepsilon(q_n)}(x_\varepsilon)) \rangle$$

$$= -\frac{1}{\varepsilon} \left(u^1_{m_\ell}(X_{r_\varepsilon(q_n)}(x_\varepsilon)) - u^1_{m_\ell}(X_{r_\varepsilon(q_n)}(x)) \right) + O(\varepsilon)$$

$$= \frac{1}{\varepsilon} \left(M^\ell_{r_\varepsilon(q_n)}(x) - M^\ell_{r(r_\varepsilon(q_n))}(x) \right) + O(\varepsilon).$$

Since $M^\ell(x)$ attains its minimum over $[\tau_\ell, q_n]$ at time $r(q_n)$ and its minimum over $[\tau_\ell, r_\varepsilon(q_n)]$ at time $r(r_\varepsilon(q_n))$, respectively, we finally get

$$|\langle \eta_{r_\varepsilon(q_n)}(\varepsilon), n(X_{r_\varepsilon(q_n)}(x_\varepsilon)) \rangle| \leq \frac{1}{|\varepsilon|} \left(M^\ell_{r_\varepsilon(q_n)}(x) - M^\ell_{r(q_n)}(x) + M^\ell_{r(r_\varepsilon(q_n))}(x) - M^\ell_{r(q_n)}(x) \right) + O(\varepsilon)$$

$$\leq \frac{2}{|\varepsilon|} \left(M^\ell_{r_\varepsilon(q_n)}(x) - M^\ell_{r(q_n)}(x) \right) + O(\varepsilon),$$

which tends to zero again by Corollary 2.14 i). □

Since for every $m \geq 0$ the coordinate mapping u_m is one to one, the set $\{\nabla u_m^i(x), i = 2, \ldots, d\}$ is linear independent for all $x \in U_m$ and by construction it is also a basis of the tangent space at x if $x \in \partial G \cap U_m$. Let $\{\bar{n}_2^m(x), \ldots, \bar{n}_d^m(x)\}$ be the Gram-Schmidt orthonormalization of $\{\nabla u_m^i(x), i = 2, \ldots, d\}$ for every $x \in U_m$ and for every m. Then, $\bar{n}^m(x) := \{n(x), \bar{n}_2^m(x), \ldots, \bar{n}_d^m(x)\}$ is an ONB of $\mathbb{R}^d$ for all $x \in U_m \cap \partial G$. We shall now extend $\hat{\eta}$ to a right-continuous process by defining $\hat{\eta}_t$ for $t \in C \cap [\tau_\ell, \tau_{\ell+1})$ in the coordinates w.r.t. the basis $\bar{n}^{m_\ell}(X_t(x))$ on $U_{m_\ell} \cap \partial G$. For that purpose it suffices to define

$$\langle \hat{\eta}_t, \nabla u_{m_\ell}^1(X_t(x)) \rangle = \langle \hat{\eta}_t, n(X_t(x)) \rangle := 0,$$

and for $i = 2, \ldots, d$,

$$\langle \hat{\eta}_t, \nabla u_{m_\ell}^i(X_t(x)) \rangle := \nabla u_{m_\ell}^i(X_{\tau_\ell}(x)) \cdot \hat{\eta}_{\tau_\ell} + \int_{[\tau_\ell, t) \setminus C} \nabla b_{m_\ell}^i(X_r(x)) \cdot \hat{\eta}_r \, dr$$

$$+ \sum_{k=1}^d \int_{[\tau_\ell, t) \setminus C} \nabla \sigma_{m_\ell}^{ik}(X_r(x)) \cdot \hat{\eta}_r \, dw_r^k.$$

Remark 2.16. This definition leads in fact to a right-continuous extension of $\hat{\eta}$. On one hand we have $\langle \hat{\eta}_{r(q_n)}, n(X_{r(q_n)}(x)) \rangle = 0$ for every n by Lemma 2.15 i), and on the other hand we have for all $t \in [\tau_\ell, \tau_{\ell+1})$ and $i = 2, \ldots, d$

$$\langle \hat{\eta}_t, \nabla u_{m_\ell}^i(X_t(x)) \rangle = \nabla u_{m_\ell}^i(X_{\tau_\ell}(x)) \cdot \hat{\eta}_{\tau_\ell} + \int_{\tau_\ell}^t \nabla b_{m_\ell}^i(X_r(x)) \cdot \hat{\eta}_r \, dr$$

$$+ \sum_{k=1}^d \int_{\tau_\ell}^t \nabla \sigma_{m_\ell}^{ik}(X_r(x)) \cdot \hat{\eta}_r \, dw_r^k. \tag{2.21}$$

Indeed, for $t \in C$ this is just the definition and for $t \in [\tau_\ell, \tau_{\ell+1}) \setminus C$ we have by Taylor's formula and (2.7) that

$$\langle \hat{\eta}_t, \nabla u_{m_\ell}^i(X_t(x)) \rangle = \lim_{l \to \infty} \frac{1}{\varepsilon_{\nu_l}} \left(u_{m_\ell}^i(X_t(x_{\varepsilon_{\nu_l}})) - u_{m_\ell}^i(X_t(x)) \right)$$

$$= \lim_{l \to \infty} \frac{1}{\varepsilon_{\nu_l}} \left(u_{m_\ell}^i(X_{\tau_\ell}(x_{\varepsilon_{\nu_l}})) - u_{m_\ell}^i(X_{\tau_\ell}(x)) + \int_{\tau_\ell}^t \left(b_{m_\ell}^i(X_{\tau_\ell}(x_{\varepsilon_{\nu_l}})) - b_{m_\ell}^i(X_{\tau_\ell}(x)) \right) dr \right.$$

$$\left. + \sum_{k=1}^d \int_{\tau_\ell}^t \left(\sigma_{m_\ell}^{ik}(X_{\tau_\ell}(x_{\varepsilon_{\nu_l}})) - \sigma_{m_\ell}^{ik}(X_{\tau_\ell}(x)) \right) dw_r^k \right),$$

and this converges to the right hand side of (2.21) by a similar argument as in the proof of Lemma 2.17 below. In particular, this argument does not depend on the value of $\hat{\eta}_t$ for $t \in C$ since C has zero Lebesgue measure.

Let for all $x \in U_m$, $m \geq 0$ and $\eta \in \mathbb{R}^d$

$$\tilde{\Pi}_x^m(\eta) := \sum_{k=2}^d \langle \eta, \bar{n}_k^m(x) \rangle \bar{n}_k^m(x), \tag{2.22}$$

2.3 Proof of the Main Result

so that obviously

$$\tilde{\Pi}_x^m(\eta) = \pi_x(\eta), \quad \forall x \in \partial G \cap U_m, \forall \eta \in \mathbb{R}^d.$$

For later use we prove now uniform convergence of $\tilde{\Pi}_{X_t(x_\varepsilon)}^{m_\ell}(\eta_t(\varepsilon))$ to $\tilde{\Pi}_{X_t(x)}^{m_\ell}(\hat{\eta}_t)$ along the chosen subsequence. The proof is based on the fact that there are no local time terms appearing in equation (2.7) for $u_{m_\ell}^i$, $i = 2, \ldots, d$. In particular, note that $\tilde{\Pi}_{X_t(x)}^{m_\ell}(\hat{\eta}_t)$ is not the same as $Q \cdot O_t \cdot \hat{\eta}_t$. Later we will identify that process with Y_t^2 appearing in Theorem 2.5, which does depend on the local time. Both processes do only coincide for $t \in [\tau_\ell, \tau_{\ell+1}] \cap C$.

Lemma 2.17. *For every $\ell \geq 0$ let $\Delta_{\tau_{\ell+1}} > 0$ be as in Remark 2.9 such that $\tilde{\Pi}_{X_s(x_\varepsilon)}^{m_\ell}(\eta_s(\varepsilon))$ is well defined for all $s \in [\tau_\ell, \tau_{\ell+1}]$ and all $0 < |\varepsilon| < \Delta_{\tau_{\ell+1}}$. Then,*

$$\sup_{s \in [\tau_\ell, \tau_{\ell+1}]} \left| \tilde{\Pi}_{X_s(x_{\varepsilon_{\nu_l}})}^{m_\ell}(\eta_s(\varepsilon_{\nu_l})) - \tilde{\Pi}_{X_s(x)}^{m_\ell}(\hat{\eta}_s) \right| \mathbb{1}_{\{0 < |\varepsilon| < \Delta_{\tau_{\ell+1}}\}} \to 0 \quad \text{in } L^p, \, p > 1 \text{ as } l \to \infty,$$

in particular for every $q_n > \inf C$ contained in $[\tau_\ell, \tau_{\ell+1}]$,

$$\left| \tilde{\Pi}_{X_{r(q_n)}(x_{\varepsilon_{\nu_l}})}^{m_\ell}(\eta_{r(q_n)}(\varepsilon_{\nu_l})) - \tilde{\Pi}_{X_{r(q_n)}(x)}^{m_\ell}(\hat{\eta}_{r(q_n)}) \right| \mathbb{1}_{\{0 < |\varepsilon| < \Delta_{\tau_{\ell+1}}\}} \to 0 \quad \text{in } L^p, \, p > 1 \text{ as } l \to \infty.$$

Proof. Since every function $\bar{n}_k^m$ is continuous on U_m, it suffices by Proposition 2.8 to show that

$$\sup_{s \in [\tau_\ell, \tau_{\ell+1}]} \left| \tilde{\Pi}_{X_s(x)}^{m_\ell}(\eta_s(\varepsilon_{\nu_l})) - \tilde{\Pi}_{X_s(x)}^{m_\ell}(\hat{\eta}_s) \right| \mathbb{1}_{\{0 < |\varepsilon| < \Delta_{\tau_{\ell+1}}\}} \to 0 \quad \text{in } L^p, \, p > 2, \text{ as } l \to \infty,$$

and for this it is enough to prove that for every $i \in \{2, \ldots, d\}$,

$$\sup_{s \in [\tau_\ell, \tau_{\ell+1}]} \left| \langle \eta_s(\varepsilon_{\nu_l}), \nabla u_{m_\ell}^i(X_s(x)) \rangle - \langle \hat{\eta}_s, \nabla u_{m_\ell}^i(X_s(x)) \rangle \right| \mathbb{1}_{\{0 < |\varepsilon| < \Delta_{\tau_{\ell+1}}\}} \to 0 \quad \text{in } L^p \text{ as } l \to \infty.$$

As before we use Taylor's formula and (2.7) to obtain

$$\begin{aligned}
\langle \eta_s(\varepsilon), \nabla u_{m_\ell}^i(X_s(x)) \rangle =& \frac{1}{\varepsilon} \left(u_{m_\ell}^i(X_s(x_\varepsilon)) - u_{m_\ell}^i(X_s(x)) \right) + O(\varepsilon) \\
=& \frac{1}{\varepsilon} \left(u_{m_\ell}^i(X_{\tau_\ell}(x_\varepsilon)) - u_{m_\ell}^i(X_{\tau_\ell}(x)) + \int_{\tau_\ell}^s \left(b_{m_\ell}^i(X_r(x_\varepsilon)) - b_{m_\ell}^i(X_r(x)) \right) dr \right. \\
& \left. + \sum_{k=1}^d \int_{\tau_\ell}^s \left(\sigma_{m_\ell}^{ik}(X_r(x_\varepsilon)) - \sigma_{m_\ell}^{ik}(X_r(x)) \right) dw_r^k \right) + O(\varepsilon) \\
=& \nabla u_{m_\ell}^i(X_{\tau_\ell}(x)) \cdot \eta_{\tau_\ell}(\varepsilon) + \int_{\tau_\ell}^s \int_0^1 \nabla b_{m_\ell}^i(X_r^{\alpha, \varepsilon}) \cdot \eta_r(\varepsilon) \, d\alpha \, dr \\
& + \sum_{k=1}^d \int_{\tau_\ell}^s \int_0^1 \nabla \sigma_{m_\ell}^{ik}(X_r^{\alpha, \varepsilon}) \cdot \eta_r(\varepsilon) \, d\alpha \, dw_r^k + O(\varepsilon), \quad (2.23)
\end{aligned}$$

where as before $X_r^{\alpha,\varepsilon} := \alpha X_r(x_\varepsilon) + (1-\alpha)X_r(x)$, $\alpha \in [0,1]$. Recall the definition of $\hat{\tau}_\ell(x_\varepsilon)$ in Remark 2.9. Comparing (2.21) and (2.23) leads to

$$\mathbb{E}\left[\sup_{s\in[\tau_\ell,\tau_{\ell+1}]}\left|\langle \eta_s(\varepsilon), \nabla u_{m_\ell}^i(X_s(x))\rangle - \langle \hat{\eta}_s, \nabla u_{m_\ell}^i(X_s(x))\rangle\right|^p \mathbb{1}_{\{0<|\varepsilon|<\Delta_{\tau_{\ell+1}}\}}\right]$$

$$\leq c_1 \mathbb{E}\left[\|\nabla u_{m_\ell}^i(X_{\tau_\ell}(x))\|^p \|\eta_{\tau_\ell}(\varepsilon) - \hat{\eta}_{\tau_\ell}\|^p\right]$$

$$+ c_1 \mathbb{E}\left[\int_{\tau_\ell}^{\tau_{\ell+1}} \left|\int_0^1 \nabla b_{m_\ell}^i(X_r^{\alpha,\varepsilon})\,d\alpha \cdot \eta_r(\varepsilon) - \nabla b_{m_\ell}^i(X_r(x))\cdot \hat{\eta}_r\right|^p dr\, \mathbb{1}_{\{0<|\varepsilon|<\Delta_{\tau_{\ell+1}}\}}\right]$$

$$+ c_1 \sum_{k=1}^d \mathbb{E}\left[\sup_{s\in[\tau_\ell,\hat{\tau}_\ell(x_\varepsilon)]}\left|\int_{\tau_\ell}^s \left(\int_0^1 \nabla \sigma_{m_\ell}^{ik}(X_r^{\alpha,\varepsilon})\,d\alpha \cdot \eta_r(\varepsilon) - \nabla \sigma_{m_\ell}^{ik}(X_r(x))\cdot \hat{\eta}_r\right) dw_r^k\right|^p\right] + O(\varepsilon).$$

Using Burkholder's inequality the third term can be estimated by

$$c_2 \sum_{k=1}^d \mathbb{E}\left[\int_{\tau_\ell}^{\hat{\tau}_\ell(x_\varepsilon)}\left|\int_0^1 \nabla \sigma_{m_\ell}^{ik}(X_r^{\alpha,\varepsilon})\,d\alpha \cdot \eta_r(\varepsilon) - \nabla \sigma_{m_\ell}^{ik}(X_r(x))\cdot \hat{\eta}_r\right|^p dr\right].$$

Finally, we obtain by Proposition 2.8,

$$\mathbb{E}\left[\sup_{s\in[\tau_\ell,\tau_{\ell+1}]}\left|\langle \eta_s(\varepsilon), \nabla u_{m_\ell}^i(X_s(x))\rangle - \langle \hat{\eta}_s, \nabla u_{m_\ell}^i(X_s(x))\rangle\right|^p \mathbb{1}_{\{0<|\varepsilon|<\Delta_{\tau_{\ell+1}}\}}\right]$$

$$\leq c_3 \mathbb{E}\left[\|\nabla u_{m_\ell}^i(X_{\tau_\ell}(x))\|^p \|\eta_{\tau_\ell}(\varepsilon) - \hat{\eta}_{\tau_\ell}\|^p\right]$$

$$+ c_3 \mathbb{E}\left[\int_{\tau_\ell}^{\tau_{\ell+1}}\left|\int_0^1 \nabla b_{m_\ell}^i(X_r^{\alpha,\varepsilon})\,d\alpha - \nabla b_{m_\ell}^i(X_r(x))\right|^p dr\, \mathbb{1}_{\{0<|\varepsilon|<\Delta_{\tau_{\ell+1}}\}} e^{c_4(T+l_T(x)+l_T(x_\varepsilon))}\right]$$

$$+ c_3 \sum_{k=1}^d \mathbb{E}\left[\int_{\tau_\ell}^{\hat{\tau}_\ell(x_\varepsilon)}\left|\int_0^1 \nabla \sigma_{m_\ell}^{ik}(X_r^{\alpha,\varepsilon})\,d\alpha - \nabla \sigma_{m_\ell}^{ik}(X_r(x))\right|^p dr \cdot e^{c_4(T+l_T(x)+l_T(x_\varepsilon))}\right]$$

$$+ c_3 \mathbb{E}\left[\int_{\tau_\ell}^{\tau_{\ell+1}} \|\eta_r(\varepsilon) - \hat{\eta}_r\|^p dr\right] + O(\varepsilon),$$

which converges to zero along (ε_{ν_l}) by dominated convergence, since $\eta_{\tau_\ell}(\varepsilon_{\nu_l}) \to \hat{\eta}_{\tau_\ell}$, $X_r^{\alpha,\varepsilon_{\nu_l}} \to X_r(x)$ uniformly in $r \in [0,T]$, $\nabla b_{m_\ell}^i$ and $\nabla \sigma_{m_\ell}^{ik}$ are continuous and $\eta_r(\varepsilon_{\nu_l})$ converges to $\hat{\eta}_r$ uniformly in $r \in A_n$ for every n. $\square$

A Characterizing Equation for the Derivatives

We have proven so far that $\eta_t(\varepsilon)$ converges along a subsequence to $\hat{\eta}_t$ a.s. for every t. In order to show differentiability we still need to show that $\hat{\eta}$ does not depend on the chosen subsequence $(\varepsilon_{\nu_l})_l$. To that aim we shall establish a system of SDE-like equations, which admits a unique solution and which is solved by $Y_t := O_t \cdot \hat{\eta}_t$, $t \in [0,T]$, O_t denoting the moving frame defined in Section 2.2.2. We shall proceed similarly to Section 4 in [1] (see also Section V.6 in [42]).

2.3 Proof of the Main Result

At first we shall derive an equation for $Y_t(\varepsilon) := O_t \cdot \eta_t(\varepsilon)$, $t \in [0,T]$. Let the rows of O_t be denoted by $n_t^k = n^k(X_t(x))$, $k = 1, \ldots d$. Then, we obtain by the chain rule that for every t

$$\frac{1}{\varepsilon}[b(X_t(x_\varepsilon)) - b(X_t(x))] = \sum_{k=1}^d \int_0^1 D_{n_t^k} b(X_t^{\alpha,\varepsilon}) \cdot \langle \eta_t(\varepsilon), n_t^k \rangle \, d\alpha = \int_0^1 Db(X_t^{\alpha,\varepsilon}) \, d\alpha \cdot O_t^{-1} \cdot Y_t(\varepsilon).$$

By Itô's integration by parts formula we have

$$\begin{aligned}
dY_t(\varepsilon) =& O_t \cdot d\eta_t(\varepsilon) + dO_t \cdot \eta_t(\varepsilon) \\
=& O_t \cdot \frac{1}{\varepsilon}[b(X_t(x_\varepsilon)) - b(X_t(x))] \, dt + O_t \cdot \frac{1}{\varepsilon}[n(X_t(x_\varepsilon))dl_t(x_\varepsilon) - n(X_t(x))dl_t(x)] + dO_t \cdot \eta_t \\
=& \left[O_t \cdot \int_0^1 Db(X_t^{\alpha,\varepsilon}) \, d\alpha \cdot O_t^{-1} + \beta(X_t(x)) \cdot O_t^{-1} \right] \cdot Y_t(\varepsilon) \, dt \\
& + \sum_{k=1}^d \alpha_k(X_t(x)) \cdot O_t^{-1} \cdot Y_t(\varepsilon) \, dw_t^k + \gamma(X_t(x)) \cdot O_t^{-1} \cdot Y_t(\varepsilon) \, dl_t(x) \\
& + O_t \cdot \frac{1}{\varepsilon}[n(X_t(x_\varepsilon))dl_t(x_\varepsilon) - n(X_t(x))dl_t(x)],
\end{aligned}$$

with coefficient functions α_k and β and γ as in (2.2). Let $P = \text{diag}(e^1)$ and $Q = \text{Id} - P$ and set

$$Y_t^{1,\varepsilon} = P \cdot Y_t(\varepsilon) \quad \text{and} \quad Y_t^{2,\varepsilon} = Q \cdot Y_t(\varepsilon)$$

to decompose the space $\mathbb{R}^d$ into the direct sum $\text{Im } P \oplus \text{Ker } P$. We define the coefficients c and d_ε to be such that

$$\sum_{k=1}^d \begin{pmatrix} c_k^1(t) & c_k^2(t) \\ c_k^3(t) & c_k^4(t) \end{pmatrix} dw_t^k + \begin{pmatrix} d_\varepsilon^1(t) & d_\varepsilon^2(t) \\ d_\varepsilon^3(t) & d_\varepsilon^4(t) \end{pmatrix} dt$$

$$= \sum_{k=1}^d \alpha_k(X_t(x)) \cdot O_t^{-1} \, dw_t^k + \left[O_t \cdot \int_0^1 Db(X_t^{\alpha,\varepsilon}) \, d\alpha \cdot O_t^{-1} + \beta(X_t(x)) \cdot O_t^{-1} \right] dt.$$

As before let $t \in [0,T]\backslash C$ and let n be such that $t \in A_n$. Then there exists $\Delta_n > 0$ such that a.s. $l_{q_n}(x) = l_{q_n}(x_\varepsilon) = 0$ if $q_n < \inf C$ and both of them are strictly positive if $q_n > \inf C$ for all $0 < |\varepsilon| < \Delta_n$. For such ε we get

$$Y_t^{1,\varepsilon} = Y_0^{1,\varepsilon} + \sum_{k=1}^d \int_0^t \left(c_k^1(s) Y_s^{1,\varepsilon} + c_k^2(s) Y_s^{2,\varepsilon} \right) dw_s^k + \int_0^t \left(d_\varepsilon^1(s) Y_s^{1,\varepsilon} + d_\varepsilon^2(s) Y_s^{2,\varepsilon} \right) ds, \qquad (2.24)$$

if $t < \inf C$ and

$$Y_t^{1,\varepsilon} = Y_{r(t)}^{1,\varepsilon} + \sum_{k=1}^d \int_{r(t)}^t \left(c_k^1(s) Y_s^{1,\varepsilon} + c_k^2(s) Y_s^{2,\varepsilon} \right) dw_s^k + \int_{r(t)}^t \left(d_\varepsilon^1(s) Y_s^{1,\varepsilon} + d_\varepsilon^2(s) Y_s^{2,\varepsilon} \right) ds + R_t(\varepsilon),$$

(2.25)

if $t \geq \inf C$, where $R_t(\varepsilon) := P \cdot O_t \cdot R_{q_n}(x_\varepsilon)$ with $R_{q_n}(x_\varepsilon)$ as in (2.19).

In order to compute the corresponding equation for $Y^{2,\varepsilon}$, we use again Taylor's formula to obtain for every $k \in \{2,\ldots,d\}$

$$n^k(X_t(x)) \cdot \tfrac{1}{\varepsilon}\left[n(X_t(x_\varepsilon))\,dl_t(x_\varepsilon) - n(X_t(x))\,dl_t(x)\right]$$
$$=\tfrac{1}{\varepsilon}\left[(n^k(X_t(x)) - n^k(X_t(x_\varepsilon))) \cdot n(X_t(x_\varepsilon))\,dl_t(x_\varepsilon)\right]$$
$$= -\eta_t(\varepsilon)^* \cdot Dn^k(X_t(x_\varepsilon))^* \cdot n(X_t(x_\varepsilon))\,dl_t(x_\varepsilon) + O(\varepsilon)$$
$$=\eta_t(\varepsilon)^* \cdot Dn(X_t(x_\varepsilon))^* \cdot n^k(X_t(x_\varepsilon))^*\,dl_t(x_\varepsilon) + O(\varepsilon)$$
$$=n^k(X_t(x_\varepsilon)) \cdot Dn(X_t(x_\varepsilon)) \cdot \eta_t(\varepsilon)\,dl_t(x_\varepsilon) + O(\varepsilon)$$
$$=n^k(X_t(x_\varepsilon)) \cdot Dn(X_t(x_\varepsilon)) \cdot O_t^{-1} \cdot Y_t(\varepsilon)\,dl_t(x_\varepsilon) + O(\varepsilon).$$

Hence,

$$Q \cdot O_t \cdot \tfrac{1}{\varepsilon}\left[n(X_t(x_\varepsilon))dl_t(x_\varepsilon) - n(X_t(x))dl_t(x)\right] = \Phi_\varepsilon(t) \cdot Y_t(\varepsilon)\,dl_t(x_\varepsilon) + O(\varepsilon)$$
$$= \left[\Phi^1_\varepsilon(t) \cdot Y^{1,\varepsilon}_t + \Phi^2_\varepsilon(t) \cdot Y^{2,\varepsilon}_t\right] + O(\varepsilon),$$

where

$$\Phi_\varepsilon(t) := Q \cdot O(X_t(x_\varepsilon)) \cdot Dn(X_t(x_\varepsilon)) \cdot O_t^{-1} \quad \text{and} \quad \Phi^1_\varepsilon(t) := \Phi_\varepsilon(t) \cdot P, \quad \Phi^2_\varepsilon(t) := \Phi_\varepsilon(t) \cdot Q.$$

Finally, we obtain the following equation for $Y^{2,\varepsilon}$:

$$Y^{2,\varepsilon}_t = Y^{2,\varepsilon}_0 + \sum_{k=1}^d \int_0^t \left(c^3_k(s)Y^{1,\varepsilon}_s + c^4_k(s)Y^{2,\varepsilon}_s\right) dw^k_s + \int_0^t \left(d^3_\varepsilon(s)Y^{1,\varepsilon}_s + d^4_\varepsilon(s)Y^{2,\varepsilon}_s\right) ds$$
$$+ \int_0^t \left(\Phi^1_\varepsilon(s)Y^{1,\varepsilon}_s + \Phi^2_\varepsilon(s)Y^{2,\varepsilon}_s\right) dl_s(x_\varepsilon) + \int_0^t \left(\gamma^1(s)Y^{1,\varepsilon}_s + \gamma^2(s)Y^{2,\varepsilon}_s\right) dl_s(x) + O(\varepsilon),$$
(2.26)

with $\gamma^1(t) := \gamma(X_t(x)) \cdot O_t^{-1} \cdot P$ and $\gamma^2(t) := \gamma(X_t(x)) \cdot O_t^{-1} \cdot Q$.

Setting $Y_t = O_t \cdot \hat\eta_t$ and $Y^1_t = P \cdot Y_t$, $Y^2_t = Q \cdot Y_t$, $t \in [0,T]$, the next step is to prove the following

Proposition 2.18. *For every $t \in [0,T]$, Y_t satisfies the equations in Theorem 2.5.*

Obviously, from the second equation in Theorem 2.5 it follows that Y^2_t is a continuous semimartingale in t, which in particular implies that the mapping $t \mapsto \pi_{X_{r(q_n)}(x)}(\hat\eta_t)$ is continuous at time $t = r(q_n)$ for every n. Thus, the proof of Theorem 2.1 is complete once we have shown Proposition 2.18 and that system in Theorem 2.5 admits a unique solution. First, we prove two preparing lemmas.

Lemma 2.19. *For every $t > 0$,*

i) $\int_0^t \Phi^1_\varepsilon(s) Y^{1,\varepsilon}_s\,dl_s(x_\varepsilon) \to 0$ *in* L^2 *as* $\varepsilon \to 0$,

ii) $\int_0^t \gamma^1(s) Y^{1,\varepsilon}_s\,dl_s(x) \to 0$ *in* L^2 *as* $\varepsilon \to 0$.

2.3 Proof of the Main Result

Proof. Since Φ_ε^1 is uniformly bounded, we get

$$\left\| \int_0^t \Phi_\varepsilon^1(s)\, Y_s^{1,\varepsilon}\, dl_s(x_\varepsilon) \right\| \leq c_1 \int_0^t |\langle \eta_s(\varepsilon), n(X_s(x))\rangle|\, dl_s(x_\varepsilon)$$

$$\leq c_1 \int_0^t |\langle \eta_s(\varepsilon), n(X_s(x_\varepsilon))\rangle|\, dl_s(x_\varepsilon) + c_1 \int_0^t |\langle \eta_s(\varepsilon), n(X_s(x_\varepsilon)) - n(X_s(x))\rangle|\, dl_s(x_\varepsilon), \quad (2.27)$$

where the second term tends to zero by Proposition 2.8. Let now $\sigma_s^\varepsilon := \inf\{r : l_r(x_\varepsilon) \geq s\}$ be the left-continuous inverse of $l(x_\varepsilon)$, i.e. for every s, σ_s^ε is the left endpoint of an excursion interval of $X(x_\varepsilon)$. In particular, for every fixed s, $\sigma_s^\varepsilon = r_\varepsilon(q_n)$ for some q_n depending on s if ε is small. Thus, for every positive M we obtain by Lemma 2.15 ii)

$$\mathbb{E}\left[\int_0^M |\langle \eta_{\sigma_s^\varepsilon}(\varepsilon), n(X_{\sigma_s^\varepsilon}(x_\varepsilon))\rangle|^2\, ds\right] = \int_0^M \mathbb{E}\left[\left|\langle \eta_{r_\varepsilon(q_n)}(\varepsilon), n(X_{r_\varepsilon(q_n)}(x_\varepsilon))\rangle\right|^2\right] ds \to 0, \quad \text{as } \varepsilon \to 0. \tag{2.28}$$

We show now that also the first term in (2.27) tends to zero. On one hand, we have by the change of variables formula for an arbitrary $M > 0$

$$\int_0^t |\langle \eta_s(\varepsilon), n(X_s(x_\varepsilon))\rangle|\, dl_s(x_\varepsilon)\, \mathbb{1}_{\{l_t(x_\varepsilon) \leq M\}} = \int_0^{l_t(x_\varepsilon)} |\langle \eta_{\sigma_s^\varepsilon}(\varepsilon), n(X_{\sigma_s^\varepsilon}(x_\varepsilon))\rangle|\, ds\, \mathbb{1}_{\{l_t(x_\varepsilon) \leq M\}}$$

$$\leq \int_0^M |\langle \eta_{\sigma_s^\varepsilon}(\varepsilon), n(X_{\sigma_s^\varepsilon}(x_\varepsilon))\rangle|\, ds,$$

which converges to zero in L^2 by (2.28). On the other hand, using the Cauchy-Schwarz inequality and Proposition 2.8 we get

$$\mathbb{E}\left[\left(\int_0^t |\langle \eta_s(\varepsilon), n(X_s(x_\varepsilon))\rangle|\, dl_s(x_\varepsilon)\right)^2 \mathbb{1}_{\{l_t(x_\varepsilon) > M\}}\right] \leq \mathbb{E}\left[\exp(c_2(t + l_t(x) + l_t(x_\varepsilon)))\, l_t(x_\varepsilon)^4\right]^{1/2}$$

$$\times \mathbb{P}[l_t(x_\varepsilon) > M]^{1/2}.$$

Hence,

$$\limsup_{\varepsilon \to 0} \mathbb{E}\left[\left(\int_0^t |\langle \eta_s(\varepsilon), n(X_s(x_\varepsilon))\rangle|\, dl_s(x_\varepsilon)\right)^2\right] \leq \mathbb{E}[\exp(c_4(l_t(x) + t))\, l_t(x)^4]^{1/2}\, \mathbb{P}[l_t(x) > M]^{1/2}.$$

We let M tend to infinity and obtain that i) holds. ii) follows by an analogous, simpler proceeding. $\square$

Lemma 2.20. *For every $t > 0$,*

i) $\int_0^t \Phi_{\varepsilon_{\nu_l}}^2(s)\, Y_s^{2,\varepsilon_{\nu_l}}\, dl_s(x_{\varepsilon_{\nu_l}}) \to \int_0^t \Phi^2(s)\, Y_s^2\, dl_s(x)$ *in L^2 as $l \to \infty$,*

ii) $\int_0^t \gamma^2(s)\, Y_s^{2,\varepsilon_{\nu_l}}\, dl_s(x) \to \int_0^t \gamma^2(s)\, Y_s^2\, dl_s(x)$ *in L^2 as $l \to \infty$.*

Proof. Again we only prove i). Since

$$\int_0^t \Phi_\varepsilon^2(s) Y_s^{2,\varepsilon} dl_s(x_\varepsilon) - \Phi^2(s) Y_s^2 dl_s(x) = \sum_{\ell:\tau_\ell<t} \int_{\tau_\ell}^{\tau_{\ell+1}\wedge t} \Phi_\varepsilon^2(s) Y_s^{2,\varepsilon} dl_s(x_\varepsilon) - \Phi^2(s) Y_s^2 dl_s(x),$$

and the total number of summands on the right hand side is finite a.s., it suffices to show that for every ℓ,

$$\int_{\tau_\ell}^{\tau_{\ell+1}} \Phi_\varepsilon^2(s) Y_s^{2,\varepsilon} dl_s(x_\varepsilon) - \Phi^2(s) Y_s^2 dl_s(x) \to 0 \quad \text{in } L^2 \text{ along } \varepsilon_{\nu_l}.$$

Recall the definition of $\tilde{\Pi}_x^m$ in (2.22). By construction we have for $s \in [\tau_\ell, \tau_{\ell+1}]$,

$$\Phi^2(s) Y_s^2 dl_s(x) = \Phi^2(s) \cdot Q \cdot O_s \cdot \hat{\eta}_s dl_s(x) = \Phi^2(s) \cdot Q \cdot O_s \cdot \pi_{X_s(x)}(\hat{\eta}_s) dl_s(x)$$
$$= \Phi^2(s) \cdot Q \cdot O_s \cdot \tilde{\Pi}_{X_s(x)}^{m_\ell}(\hat{\eta}_s) dl_s(x),$$

and analogously, for sufficiently small ε,

$$\Phi_\varepsilon^2(s) Y_s^{2,\varepsilon} dl_s(x_\varepsilon)$$
$$= \left\{ \Phi_\varepsilon^2(s) \cdot Q \cdot O(X_s(x_\varepsilon)) \cdot \eta_s(\varepsilon) + \Phi_\varepsilon^2(s) \cdot Q \cdot [O(X_s(x)) - O(X_s(x_\varepsilon))] \cdot \eta_s(\varepsilon) \right\} dl_s(x_\varepsilon)$$
$$= \left\{ \Phi_\varepsilon^2(s) \cdot Q \cdot O(X_s(x_\varepsilon)) \cdot \tilde{\Pi}_{X_s(x_\varepsilon)}^{m_\ell}(\eta_s(\varepsilon)) + \Phi_\varepsilon^2(s) \cdot Q \cdot [O(X_s(x)) - O(X_s(x_\varepsilon))] \cdot \eta_s(\varepsilon) \right\} dl_s(x_\varepsilon).$$

Hence,

$$\int_{\tau_\ell}^{\tau_{\ell+1}} \Phi_\varepsilon^2(s) Y_s^{2,\varepsilon} dl_s(x_\varepsilon) - \Phi^2(s) Y_s^2 dl_s(x)$$
$$= \int_{\tau_\ell}^{\tau_{\ell+1}} \left[\Phi_\varepsilon^2(s) \cdot Q \cdot O(X_s(x_\varepsilon)) \cdot \tilde{\Pi}_{X_s(x_\varepsilon)}^{m_\ell}(\eta_s(\varepsilon)) - \Phi^2(s) \cdot Q \cdot O_s \cdot \tilde{\Pi}_{X_s(x)}^{m_\ell}(\hat{\eta}_s) \right] dl_s(x_\varepsilon)$$
$$+ \int_{\tau_\ell}^{\tau_{\ell+1}} \Phi_\varepsilon^2(s) \cdot Q \cdot [O(X_s(x)) - O(X_s(x_\varepsilon))] \cdot \eta_s(\varepsilon) dl_s(x_\varepsilon)$$
$$+ \int_{\tau_\ell}^{\tau_{\ell+1}} \Phi^2(s) \cdot Q \cdot O_s \cdot \tilde{\Pi}_{X_s(x)}^{m_\ell}(\hat{\eta}_s) (dl_s(x_\varepsilon) - dl_s(x)).$$

The first term converges to zero in L^2 along ε_{ν_l} by Lemma 2.17 and Proposition 2.8. The second term converges also to zero in L^2 again by Proposition 2.8. Finally, the third term tends to zero by the weak convergence of $l(x_\varepsilon)$ to $l(x)$ on $[\tau_\ell, \tau_{\ell+1}]$. Note that $\tilde{\Pi}_{X_s(x)}^{m_\ell}(\hat{\eta}_s)$ is continuous in s on $[\tau_\ell, \tau_{\ell+1}]$. □

Proof of Proposition 2.18. For every $t \in [0,T]$ we have convergence of $Y_t^{1,\varepsilon}$ and $Y_t^{2,\varepsilon}$ to Y_t^1 and Y_t^2, respectively, a.s. and in L^2 along the subsequence ε_{ν_l}. Thus, it is enough to prove that the right hand sides in (2.24), (2.25) and (2.26) converge along ε_{ν_l} in L^2 to the corresponding terms in the equation describing Y^1 and Y^2. Lemma 2.15 gives convergence of $Y_{r(t)}^{1,\varepsilon}$ to zero, $R_t(\varepsilon)$ tends to zero arguing as in (2.20) and in Corollary 2.14. The convergence of the terms involving the local times follows from Lemma 2.19 and Lemma 2.20. The convergence of the remaining integral terms is clear. □

2.3 Proof of the Main Result

It remains to show uniqueness, which is carried out in the next lemma.

Proposition 2.21. *The system in Theorem (2.5) admits a unique solution.*

Proof. Let $(U_t)_{t\geq 0}$ be the unique solution of the matrix-valued equation

$$U_t = \mathrm{Id} - \int_0^t U_s \cdot (\Phi^2(s) + \gamma^2(s))\, dl_s(x), \quad t \geq 0.$$

Then, introducing the stopping times $T_N := \inf\{s \geq 0 : \max(\|U_s^{-1}\|, \|U_s\|) \geq N\}$, $N \in \mathbb{N}$, we have $T_N \uparrow \infty$ a.s. as N tends to infinity. Using integration by parts we get

$$d(U_t \cdot Y_t^2) = \sum_{k=1}^d U_t \left(c_k^3(t)\, Y_t^1 + c_k^4(t)\, Y_t^2 \right) dw_t^k + U_t \left(d^3(t)\, Y_t^1 + d^4(t)\, Y_t^2 \right) dt$$
$$+ U_t\, (\Phi^2(t) + \gamma^2(t))\, Y_t^2\, dl_t(x) - U_t\, (\Phi^2(t) + \gamma^2(t))\, Y_t^2\, dl_t(x).$$

The last two terms cancel, so we can rewrite the system in Theorem 2.5 as

$$Y_t^1 = \mathbb{1}_{\{t<\inf C\}} \left(Y_0^1 + \sum_{k=1}^d \int_0^t \left(c_k^1(s)\, Y_s^1 + c_k^2(s)\, Y_s^2 \right) dw_s^k + \int_0^t \left(d^1(s)\, Y_s^1 + d^2(s)\, Y_s^2 \right) ds \right)$$
$$+ \mathbb{1}_{\{t\geq\inf C\}} \left(\sum_{k=1}^d \int_{r(t)}^t \left(c_k^1(s)\, Y_s^1 + c_k^2(s)\, Y_s^2 \right) dw_s^k + \int_{r(t)}^t \left(d^1(s)\, Y_s^1 + d^2(s)\, Y_s^2 \right) ds \right)$$

$$Y_t^2 = U_t^{-1} \cdot Y_0^2 + U_t^{-1} \cdot \sum_{k=1}^d \int_0^t U_s \cdot \left(c_k^3(s)\, Y_s^1 + c_k^4(s)\, Y_s^2 \right) dw_s^k$$
$$+ U_t^{-1} \cdot \int_0^t U_s \cdot \left(d^3(s)\, Y_s^1 + d^4(s)\, Y_s^2 \right) ds,$$

with the initial condition $Y_0^1 = P \cdot O(x) \cdot v$ and $Y_0^2 = Q \cdot O(x) \cdot v$. In order to prove the proposition, it suffices to to show existence and uniqueness of the stopped process $Y_t^N := Y_{t \wedge T_N}$, $t \in [0,T]$, since $Y_t^N \to Y_t$ a.s. for as $N \to \infty$ for every t.

We proceed as in Lemma 4.3 in [1]. Let H be the totality of $\mathbb{R}^d$-valued adapted processes (φ_t), $t \in [0,T]$, whose paths are a.s. càdlàg and which satisfy $\sup_{t \in [0,T]} \mathbb{E}[\|\varphi_t\|^2] < \infty$. On H we introduce the norm

$$\|\varphi\|_H = \sup_{t \in [0,T]} \mathbb{E}\left[\|\varphi_t\|^2\right]^{1/2}.$$

For any $\varphi \in H$ we define the process $I(\varphi)$ by

$$I(\varphi)_t^1 = \mathbb{1}_{\{t<\inf C\}} \left(Y_0^1 + \sum_{k=1}^d \int_0^t \left(c_k^1(s)\, \varphi_s^1 + c_k^2(s)\, \varphi_s^2 \right) dw_s^k + \int_0^t \left(d^1(s)\, \varphi_s^1 + d^2(s)\, \varphi_s^2 \right) ds \right)$$
$$+ \mathbb{1}_{\{t\geq\inf C\}} \left(\sum_{k=1}^d \int_{r(t)}^t \left(c_k^1(s)\, \varphi_s^1 + c_k^2(s)\, \varphi_s^2 \right) dw_s^k + \int_{r(t)}^t \left(d^1(s)\, \varphi_s^1 + d^2(s)\, \varphi_s^2 \right) ds \right)$$

$$I(\varphi)_t^2 = U_t^{-1} \cdot Y_0^2 + U_t^{-1} \cdot \sum_{k=1}^d \int_0^t U_s \cdot \left(c_k^3(s)\, \varphi_s^1 + c_k^4(s)\, \varphi_s^2 \right) dw_s^k$$
$$+ U_t^{-1} \cdot \int_0^t U_s \cdot \left(d^3(s) \cdot \varphi_s^1 + d^4(s)\, \varphi_s^2 \right) ds,$$

if $t \leq T_N \wedge T$, and $I(\varphi)_t = I(\varphi)_{T_N}$ if $t \in [T_N, T]$. By definition of T_N, we have for fixed N that all the coefficient functions are uniformly bounded in $t \in [0, T_N]$. Hence, one can easily verify that

$$\mathbb{E}\left[\|I(\varphi)_t\|^2\right] \leq c_1 \left\{1 + \int_0^t \mathbb{E}\left[\|\varphi_r\|^2\right] dr\right\} \leq c_2 \left\{1 + \sup_{t \in [0,T]} \mathbb{E}\left[\|\varphi_t\|^2\right]\right\},$$

where the constants only depend on N and T and not on φ and t. This proves that $I(\varphi) \in H$ for every $\varphi \in H$. Similarly, one can show that for any $\varphi, \psi \in H$,

$$\mathbb{E}\left[\|I(\varphi)_t - I(\psi)_t\|^2\right] \leq c_3 \int_0^t \mathbb{E}\left[\|\varphi_s - \psi_s\|^2\right] ds \leq c_4 \sup_{t \in [0,T]} \mathbb{E}[\|\varphi_t - \psi_t\|^2],$$

and hence,

$$\|I(\varphi) - I(\psi)\|_H \leq c\|\varphi - \psi\|_H.$$

For every N, existence and uniqueness of Y^N follow now by standard arguments via Picard-iteration, the details are omitted. □

Since $T > 0$ is arbitrary, Theorem 2.1 and Theorem 2.5 follow.

2.3.4 The Neumann Condition

In this final section we prove the Neumann condition stated in Corollary 2.7. Let $x \in \partial G$. By a density argument it is sufficient to consider bounded functions f, which are continuously differentiable and have bounded derivatives. Then, for each $t > 0$ we obtain by dominated convergence and the chain rule:

$$D_{n(x)} P_t f(x) = \mathbb{E}\left[\nabla f(X_t(x)) D_{n(x)} X_t(x)\right].$$

Thus, it suffices to show $D_{n(x)} X_t(x) = 0$ for all $t \geq 0$, which is equivalent to $Y_t = 0$ for all $t \geq 0$, where $Y_t = O_t \cdot \eta_t$ with $v = n(x)$. Fix some arbitray $T > 0$. Using the same notations as in Lemma 2.21, Y_t satisfies

$$Y_t^1 = \sum_{k=1}^d \int_{r(t)}^t \left(c_k^1(s) Y_s^1 + c_k^2(s) Y_s^2\right) dw_s^k + \int_{r(t)}^t \left(d^1(s) Y_s^1 + d^2(s) Y_s^2\right) ds$$

$$Y_t^2 = U_t^{-1} \cdot \sum_{k=1}^d \int_0^t U_s \cdot \left(c_k^3(s) Y_s^1 + c_k^4(s) Y_s^2\right) dw_s^k + U_t^{-1} \cdot \int_0^t U_s \cdot \left(d^3(s) Y_s^1 + d^4(s) Y_s^2\right) ds.$$

2.3 Proof of the Main Result

Note that $\inf C = 0$ and $Y_0^2 = Q \cdot O(x) \cdot n(x) = 0$. Setting $Y_t^{1,N} = P \cdot Y_t^N$, $Y_t^{2,N} = Q \cdot Y_t^N$ and $M_t := \sum_{k=1}^d \int_0^t \left(c_k^1(s) Y_s^1 + c_k^2(s) Y_s^2 \right) dw_s^k$, we obtain by Doob's inequality for every N that

$$\begin{aligned}
\mathbb{E}\left[\sup_{t \in [0,T]} \left\| Y_t^N \right\|^2 \right] &\leq \mathbb{E}\left[\sup_{t \in [0,T]} \left\| Y_t^{1,N} \right\|^2 \right] + \left[\sup_{t \in [0,T]} \left\| Y_t^{2,N} \right\|^2 \right] \\
&\leq \mathbb{E}\left[\sup_{t \in [0,T]} \left\| M_t - M_{r(t)} \right\|^2 \right] + c_1 \int_0^T \mathbb{E}\left[\sup_{r \in [0,s]} \left\| Y_r^N \right\|^2 \right] ds \\
&\leq 2 \mathbb{E}\left[\sup_{t \in [0,T]} \left\| M_t \right\|^2 \right] + c_1 \int_0^T \mathbb{E}\left[\sup_{r \in [0,s]} \left\| Y_r^N \right\|^2 \right] ds \\
&\leq c_2 \int_0^T \mathbb{E}\left[\sup_{r \in [0,s]} \left\| Y_r^N \right\|^2 \right] ds,
\end{aligned}$$

which implies by Gronwall's Lemma that $Y_t^N = 0$ for all $t \in [0,T]$ a.s. We let N tend to infinity to obtain that $Y_t = 0$ for all $t \in [0,T]$ a.s. Since $T > 0$ is arbitrary, the claim follows.

Part II
Particle Approximation of the Wasserstein Diffusion

Chapter 3

The Approximating Particle System

3.1 Introduction and Main Results

The construction and regularity of Brownian motion or more general diffusions in Euclidean domains with reflecting boundary condition is a classical subject in probability theory, lying in the intersection of regularity theory for parabolic PDE and stochastic analysis. Starting from the early works by e.g. Fukushima [37] and Tanaka [60], the field has seen perpetual research activity, c.f. [12, 51] (and e.g. [11] for a more comprehensive list of references). Typically the reflecting process can be obtained in two ways. Either via solving a system of Skorohod SDE with local time at the boundary or via Dirichlet form methods, and under suitable smoothness assumptions on the domain and the coefficients both approaches are equivalent.

In the second part of this thesis we study a very specific singular case, in which the drift coefficients of the operator may diverge and where the equivalence of the two approaches breaks down but the process exhibits good regularity nevertheless. Our case corresponds to a Dirichlet form

$$\mathcal{E}(f,f) = \int_\Omega |\nabla f|^2 q(dx)$$

on $L^2(\Omega, dq)$ for a very specific choice of domain $\Omega \subset \mathbb{R}^N$ and measure $q(dx) = q(x)dx$ (see below). Similar but more regular variants were studied under the smoothness assumption $q \in H^{1,1}(\Omega)$ respectively $\nabla(\log q) \in L^p(\Omega, dq)$, $p > N$ in [62] and [36], where a Skorohod decomposition for the induced process still holds.

In the present work we are concerned with a diffusion, which is taking values in the closure of

$$\Sigma_N = \{x \in \mathbb{R}^N : 0 < x^1 < x^2 \cdots < X^N < 1\},$$

and which is symmetric w.r.t. the measure

$$q_N(dx) = \frac{1}{Z_\beta} \prod_{i=0}^{N} (x^{i+1} - x^i)^{\frac{\beta}{N+1}-1} dx^1 dx^2 \cdots dx^N, \qquad (3.1)$$

where $Z_\beta = (\Gamma(\beta)/(\Gamma(\beta/(N+1)))^{N+1})^{-1}$ is a normalization constant and $\beta > 0$ is some parameter. That is, we study the process $(X^N_\cdot)$ generated by the $L^2(\Sigma_N, q_N)$-closure of the quadratic

form
$$\mathcal{E}^N(f,f) = \int_{\Sigma_N} |\nabla f|^2(x) \, q_N(dx), \quad f \in C^\infty(\overline{\Sigma}_N),$$
still denoted by $\mathcal{E}^N$. Its generator $\mathcal{L}$ extends the operator $(L^N, D(L^N))$

$$L^N f(x) = (\frac{\beta}{N+1} - 1) \sum_{i=1}^{N} \left(\frac{1}{x^i - x^{i-1}} - \frac{1}{x^{i+1} - x^i} \right) \frac{\partial}{\partial x^i} f(x) + \Delta f(x) \quad \text{for } x \in \Sigma_N \quad (3.2)$$

with domain

$$D(L^N) = C^2_{Neu} = \{ f \in C^2(\overline{\Sigma}_N) \,|\, \nabla f \cdot \nu = 0 \text{ on all } (n-1)\text{-dimensional faces of } \partial \Sigma_N \},$$

and ν denoting the outward normal field on $\partial \Sigma_N$. On the level of formal Itô calculus $(L^N, D(L^N))$ corresponds to an order preserving dynamics $(X^N_\cdot) \in \overline{\Sigma}_N$ for the location of N particles in the unit interval which solves the system of coupled Skorohod SDEs

$$dx^i_t = (\frac{\beta}{N+1} - 1) \left(\frac{1}{x^i_t - x^{i-1}_t} - \frac{1}{x^{i+1}_t - x^i_t} \right) dt + \sqrt{2} dw^i_t + dl^{i-1}_t - dl^i_t, \quad i = 1, \cdots, N \quad (3.3)$$

with $x^0 = 0$, $x^{N+1} = 1$ by convention, $\{w_i\}$ are independent real Brownian motions and $\{l^i\}$ are the collision local times, i.e. satisfying

$$dl^i_t \geq 0, \quad l^i_t = \int_0^t \mathbb{1}_{\{x^i_s = x^{i+1}_s\}} dl^i_s. \quad (3.4)$$

$(X^N_\cdot)$ may thus be considered as a system of coupled two sided real Bessel processes with uniform Bessel dimension $\delta = \beta/(N+1)$. Similar to the standard real Bessel process $BES(\delta)$ with Bessel dimension $\delta < 1$, the existence of X^N, even with initial condition $X_0 = x \in \Sigma_N$, is not trivial, nor are its regularity properties.

Except for mathematical curiousity our motivation for studying this process is its relation to the Wasserstein diffusion, which is an infinite dimensional diffusion process on the space of probability measures intimately related to the Wasserstein metric [66]. In the next chapter we shall prove that the normalized empirical measure of the system (3.3) converges to the Wasserstein diffusion in the high density regime for $N \to \infty$. Hence the regularity properties of $(X^N_\cdot)$ may give an indication of the regularity of the Wasserstein diffusion, although in the present work we have not managed to obtain estimates which are uniform w.r.t. the parameter N.

Remark 3.1. For simulation the dynamics of $(X^N_\cdot)$ can be approximated by $X^{N,\epsilon}_t = \mathcal{X}^{N,\epsilon}_{\lfloor t/\epsilon^2 \rfloor}$, $t \geq 0$, where $(\mathcal{X}^{N,\epsilon}_n)_{n \geq 0}$ is the Markov chain on Σ_N with transition kernel $\mu^{N,\epsilon}(x, A) = \frac{q_N(B_\epsilon(x) \cap \Sigma_N \cap A)}{q_N(B_\epsilon(x) \cap \Sigma_N)}$. An alternative approach uses a regularized version of the formal SDE (3.3) and (3.4). For illustration we present some results by courtesy of Theresa Heeg, Bonn, for the case of $N = 4$ particles with $\beta = 10$, $\beta = 1$ and $\beta = 0.3$ respectively, at large times.

3.1 Introduction and Main Results

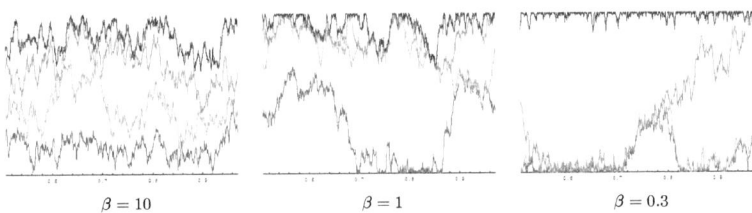

$\beta = 10$ $\qquad\qquad$ $\beta = 1$ $\qquad\qquad$ $\beta = 0.3$

We are now ready to state the main results of this chapter.

Theorem 3.2. *For $\beta < 2(N+1)$, the Dirichlet form $\mathcal{E}^N$ generates a Feller process $(X.)$ on $\overline{\Sigma}_N$, i.e. the associated transition semigroup on $L^2(\Sigma_N, q_N)$ defines a strongly continuous contraction semigroup on the subspace $C(\overline{\Sigma}_N)$ equipped with the sup-norm topology.*

In order to prove Theorem 3.2 we use some localization arguments to use the symmetries of the problem. A crucial ingredient for the justification of this strategy is the following *Markov uniqueness* result for the operator (L^N, C^2_{Neu}).

Proposition 3.3. *For $\beta < 2(N+1)$, there is at most one symmetric strongly continuous contraction semigroup $(T_t)_{t\geq 0}$ on $L^2(\Sigma_N, q_N)$ whose generator $(\mathcal{L}, \mathcal{D}(\mathcal{L}))$ extends (L^N, C^2_{Neu}).*

Hence, together with Theorem 3.2 we can state the following existence and uniqueness result.

Corollary 3.4. *The formal system of Skorohod SDEs defines via the associated martingale problem a unique diffusion process which can be started everywhere on the (closed) simplex $\overline{\Sigma}_N$.*

As for path regularity we obtain the following characterization.

Theorem 3.5. *For any starting point $x \in \overline{\Sigma}_N$, $(X_.^x) \in \overline{\Sigma}_N \subset \mathbb{R}^N$ is a Euclidean semi-martingale if and only if $\beta/(N+1) \geq 1$.*

In particular we obtain that a Skorohod decomposition of the process X^N is impossible if β is small enough. This is in sharp contrast to all aforementioned previous works. Moreover, again due to the uniqueness assertion of Proposition 3.3 the following negative result holds true.

Corollary 3.6. *If $\beta/(N+1) < 1$, the system of equations (3.3),(3.4) is ill-posed, i.e. it admits no solution in the sense of Itô calculus.*

Note that theorems 3.2 and 3.5 generalize the corresponding classical results for the family of standard real Bessel processes, which are proved in a completely different manner, c.f. [13, 56].

While the proof of Theorem 3.5 consists of a straightforward application of a regularity criterion by Fukushima [38] for Dirichlet processes, the proof of Theorem 3.2 is more involved and has four main steps. The first crucial observation is that the reference measure q_N can locally be extended to a measure $\hat{q}$ on the full Euclidean space which lies in the Muckenhoupt class

$\mathcal{A}_2$, which allows for a rich potential theory. In particular the Poincaré inequality and doubling conditions hold which imply via heat kernel estimates the regularity of the induced process on $\mathbb{R}^N$. The second step is the probabilistic piece of the localization, which corresponds to stopping the R^N-valued process at domain boundaries where the measure $\hat{q}$ is 'tame' and to show that the stopped process is again Feller. To this end we employ a version of the Wiener test for degenerate elliptic diffusions by Fabes, Jerison and Kenig [34]. The third step is to use the reflection symmetry of the problem which allows to treat the Neumann boundary condition indeed via a reflection of the extended R^N-valued process. The fourth step is to establish the Markov uniqueness for the associated martingale problem which is crucial in order to justify the identification of the processes after localization. In the proof of the Markov uniqueness we shall again depend on the nice potential theory available in Muckenhoupt classes.

In the present work we restrict ourselves to the case $\beta < 2(N+1)$ to ensure that the local extension of q_N is contained in the Muckenhoupt class $\mathcal{A}_2$. In our context this case is the more interesting one in view of the approximation result given in the next chapter. For $\beta \geq 2(N+1)$ the extension of q_N lies only in the larger Muckenhoupt class $\mathcal{A}_p$ for some $p > 2$ suffiently large and our method fails. Nevertheless, if $\beta \geq N(N+1)$ we are in the setting of [36] and the arguments there imply that the Feller property holds for (X_t^N) in our case, too.

The material presented in this chapter is contained in [10].

3.2 Dirichlet Form and Integration by Parts Formula

We start with the rigorous construction of (X_t^N), which departs from the symmetrizing measure q_N on $\overline{\Sigma}_N$ defined above in (3.1). Note that one can identify $L^p(\Sigma_N, q_N)$ with $L^p(\overline{\Sigma}_N, q_N)$, $p \geq 1$. Throughout the paper, by abuse of notation, we will also denote by q_N the density w.r.t. the Lebesgue measure, i.e. $q_N(A) = \int_A q_N(x)\,dx$ for all measurable $A \subseteq \overline{\Sigma}_N$.

For general $\beta > 0$, $N \in \mathbb{N}$, the measure q_N satisfies the 'Hamza condition', because it has a strictly positive density with locally integrable inverse, c.f. e.g. [2]. This implies that the form $\mathcal{E}^N(f,f)$ with domain $f \in C^\infty(\overline{\Sigma}_N)$ is closable on $L^2(\Sigma_N, q_N)$. The $L^2(\Sigma_N, q_N)$-closure defines a local regular Dirichlet form, still denoted by $\mathcal{E}^N$. General Dirichlet form theory asserts the existence of a Hunt diffusion (X_t^N) associated with $\mathcal{E}^N$ which can be started in q_N-almost all $x \in \overline{\Sigma}_N$ and which is understood as a generalized solution of the system (3.3),(3.4). This identification is justified by the fact that any semi-martingale solution solves via Itô's formula the martingale problem for the operator $(L^N, D(L^N))$, defined in (3.2). Next we prove an integration by parts formula for q_N.

Proposition 3.7. *Let* $u \in C^1(\overline{\Sigma}_N)$ *and* $\xi = (\xi^1, \dots, \xi^N)$ *be a vector field in* $C^1(\overline{\Sigma}_N, \mathbb{R}^N)$ *satisfying* $\langle \xi, \nu \rangle = 0$ *on* $\partial \Sigma_N$, *where* ν *denotes the outward normal field of* $\partial \Sigma_N$. *Then,*

$$\int_{\Sigma_N} \langle \nabla u, \xi \rangle q_N(dx) = -\int_{\Sigma_N} u \left[\operatorname{div}(\xi) + (\tfrac{\beta}{N+1} - 1) \sum_{i=1}^N \xi^i \left(\frac{1}{x^i - x^{i-1}} - \frac{1}{x^{i+1} - x^i} \right) \right] q_N(dx).$$

Proof. Let u and ξ be as in the statement. In a first step we shall show that

$$\int_{\Sigma_N} \operatorname{div}(u\,\xi\,q_N)\,dx = 0. \tag{3.5}$$

3.2 Dirichlet Form and Integration by Parts Formula

For k very large (compared with N) we define

$$A^k := \bigcap_{i=0}^{N} \left\{ x \in \overline{\Sigma}_N : x^i + \tfrac{1}{k} \leq x^{i+1} \right\}.$$

We apply the divergence theorem to obtain

$$\int_{A^k} \operatorname{div}(u\,\xi\,q_N)\,dx = \int_{\partial A^k} u\,\langle \xi, \nu_k \rangle q_N\,dx,$$

ν_k denoting the outward normal field of ∂A^k. Since obviously $A^k \nearrow \overline{\Sigma}_N$ as $k \to \infty$, it suffices to prove that the right hand side converges to zero as $k \to \infty$. Clearly, this term can be written as

$$\int_{\partial A^k} u\,\langle \xi, \nu_k \rangle q_N\,dx = \sum_{i=0}^{N} \int_{\partial A_i^k} u\,\langle \xi, \nu_k \rangle q_N\,dx$$

with $\partial A_i^k := A^k \cap \{x : x^i + \tfrac{1}{k} = x^{i+1}\}$. Note that for every i the outward normal on ∂A_i^k is given by the normalization of the vector $e^i - e^{i+1}$, $(e^i)_{i=1\ldots N}$ denoting the canonical basis in $\mathbb{R}^N$ and with $e^0 = e^{N+1} = 0$. Hence, for $i \in \{1, \ldots, N-1\}$

$$-\int_{\partial A_i^k} u\,\langle \xi, \nu_k \rangle q_N\,dx = \tfrac{1}{\sqrt{2}} \int_{\partial A_i^k} u(x)\left(\xi^{i+1}(x) - \xi^i(x) \right) (\tfrac{1}{k})^{\frac{\beta}{N+1} - 1} \prod_{j \neq i} (x^{j+1} - x^j)^{\frac{\beta}{N+1} - 1} dx.$$

For any $x = (x^1, \ldots, x^i, x^i + \tfrac{1}{k}, x^{i+2}, \ldots, x^N) \in \partial A_i^k$ we denote $\hat{x} = (x^1, \ldots, x^i, x^i, x^{i+2}, \ldots, x^N) \in \partial \Sigma_N \cap \{x^i = x^{i+1}\}$. Since $\xi^i(\hat{x}) - \xi^{i+1}(\hat{x}) = 0$ by the boundary condition on ξ, using the mean value theorem we get

$$\xi^{i+1}(x) - \xi^i(x) = \xi^{i+1}(x) - \xi^{i+1}(\hat{x}) + \xi^i(\hat{x}) - \xi^i(x) = \tfrac{1}{k}\left(\partial_{i+1}\xi^{i+1}(\vartheta_1) - \partial_{i+1}\xi^i(\vartheta_2) \right)$$

for some $\vartheta_1, \vartheta_2 \in [x^i, x^i + \tfrac{1}{k}]$. Thus,

$$-\int_{\partial A_i^k} u\,\langle \xi, \nu_k \rangle q_N\,dx = \tfrac{1}{\sqrt{2}} (\tfrac{1}{k})^{\frac{\beta}{N+1}} \int_{\partial A_i^k} u(x)\left(\partial_{i+1}\xi^{i+1}(\vartheta_1) - \partial_{i+1}\xi^i(\vartheta_2) \right) \prod_{j \neq i} (x^{j+1} - x^j)^{\frac{\beta}{N+1} - 1} dx.$$

For $k \to \infty$ the right hand side converges to zero, since the integral tends to

$$\int_{\partial \Sigma_N \cap \{x : x^i = x^{i+1}\}} u(x)\left(\partial_{i+1}\xi^{i+1}(x) - \partial_{i+1}\xi^i(x) \right) \prod_{j \neq i} (x^{j+1} - x^j)^{\frac{\beta}{N+1} - 1} dx < \infty.$$

For $i = 0$ and $i = N$ similar arguments apply, and (3.5) follows.

Finally, we use (3.5) to obtain

$$\int_{\Sigma_N} \langle \nabla u, \xi \rangle q_N(dx) = \int_{\Sigma_N} \operatorname{div}(u\,\xi\,q_N)\,dx - \int_{\Sigma_N} u\,\operatorname{div}(\xi)\,q_N(dx) - \int_{\Sigma_N} u\langle \xi, \nabla q_N \rangle dx$$

$$= -\int_{\Sigma_N} u\,\operatorname{div}(\xi)\,q_N(dx) - \int_{\Sigma_N} u\langle \xi, \nabla \log q_N \rangle q_N(dx),$$

and the claim follows. $\square$

Remark 3.8. Let $u \in C^1(\overline{\Sigma}_N)$ and ξ be a vector field of the the form $\xi = w\vec{\varphi}$ with $w \in C^1(\overline{\Sigma}_N)$ and $\vec{\varphi}(x) = (\varphi(x^1), \ldots, \varphi(x^N))$, $\varphi \in C^\infty([0,1])$ and $\langle \vec{\varphi}, \nu \rangle = 0$ on $\partial \Sigma_N$, in particular $\varphi_{|\partial[0,1]} = 0$. Then, the integration by parts formula in Proposition 3.7 reads

$$\int_{\Sigma_N} \langle \nabla u, \xi \rangle q_N(dx) = -\int_{\Sigma_N} u \left[w V_{N,\varphi}^\beta + \langle \nabla w, \vec{\varphi} \rangle \right] q_N(dx),$$

where

$$V_{N,\varphi}^\beta(x^1, \ldots, x^N) := (\frac{\beta}{N+1} - 1) \sum_{i=0}^{N} \frac{\varphi(x^{i+1}) - \varphi(x^i)}{x^{i+1} - x^i} + \sum_{i=1}^{N} \varphi'(x^i).$$

Let $C_{Neu}^2 = \{ f \in C^2(\overline{\Sigma}_N) : \langle \nabla f, \nu \rangle = 0$ on $\partial \Sigma_N \}$ as above with ν still denoting the outer normal field on $\partial \Sigma_N$. Then, for any $f \in C^1(\overline{\Sigma}_N)$ and $g \in C_{Neu}^2$ we apply the integration by parts formula in Proposition 3.7 for $\xi = \nabla g$ to obtain

$$\mathcal{E}^N(f, g) = -\int_{\Sigma_N} f L^N g \, q_N(dx).$$

Moreover, arguing similarly as in the proof of Proposition 3.7 we obtain that for every $g \in C_{Neu}^2$,

$$\left| \mathcal{E}^N(f, g) \right| \leq C \|f\|_{L^2(\Sigma_N, q_N)}, \quad \forall f \in D(\mathcal{E}^N).$$

In particular, C_{Neu}^2 is contained in the domain of the generator $\mathcal{L}$ associated with $\mathcal{E}^N$ and $L^N f = \mathcal{L} f$ for all $f \in C_{Neu}^2$. Hence, the process X^N is the formal solution to the system (3.3),(3.4).

3.3 Feller Property

This section is devoted to the proof of Theorem 3.2. Let from now on $\beta < 2(N+1)$. We define

$$\Omega_N := \Omega_N^\delta := \overline{\Sigma}_N \cap \{ x \in \mathbb{R}^N : x_N \leq 1 - \delta \},$$

for some positive small $\delta < [2(N+2)]^{-1}$ and the weight function

$$\hat{q}(x) := \hat{q}_{N,\delta}(x) := \begin{cases} q_N(x) & \text{if } x \in \Omega_N, \\ \frac{1}{Z_\beta} \delta^{\frac{\beta}{N+1} - 1} \prod_{i=1}^{N} (x^i - x^{i-1})^{\frac{\beta}{N+1} - 1} & \text{if } x \in S_1 \backslash \Omega_N, \end{cases} \quad (3.6)$$

where

$$S_1 := \{ x \in \mathbb{R}^N : 0 \leq x^1 \leq x^2 \leq \ldots \leq x^N \}.$$

We want to extend the weight function $\hat{q}$ to the whole $\mathbb{R}^N$. To do this we introduce the mapping

$$T : \mathbb{R}^N \to S_1 \quad x \mapsto (|x^{(1)}|, \ldots, |x^{(N)}|), \quad (3.7)$$

where $(1), \ldots, (N)$ denotes the permutation of $1, \ldots, N$ such that

$$|x^{(1)}| \leq \ldots \leq |x^{(N)}|.$$

3.3 Feller Property

The extension of $\hat{q}$ on $\mathbb{R}^N$ is now defined via $\hat{q}(x) = \hat{q}(Tx)$, $x \in \mathbb{R}^N$. Again we will also denote by $\hat{q}$ the induced measure on $\mathbb{R}^N$. Consider the $L^2(\mathbb{R}^N, \hat{q})$-closure of

$$\hat{\mathcal{E}}^{N,a}(f,f) = \int_{\mathbb{R}^N} \langle a\nabla f, a\nabla f(x) \rangle \, \hat{q}(dx), \quad f \in C_c^\infty(\mathbb{R}^N) \tag{3.8}$$

still denoted by $\hat{\mathcal{E}}^{N,a}$, for a measurable field $x \mapsto a(x) \in \mathbb{R}^{N \times N}$ on $\mathbb{R}^N$ satisfying

$$\frac{1}{c} \cdot E_N \leq a(x)^t \cdot a(x) \leq c \cdot E_N \tag{3.9}$$

in the sense of non-negative definite matrices. Here E_N denotes the identity matrix and a^t the transposition of a. Let $(Y_t)_{t \geq 0} = (Y_t^{N,a})_{t \geq 0}$ be the associated symmetric Hunt process on $\mathbb{R}^N$, starting from the invariant distribution $\hat{q}$. Finally, we denote by $(Q_t)_t$ the transition semigroup of Y.

3.3.1 Feller Properties of Y

In the sequel we will denote by $C_0(\mathbb{R}^N)$ the space of continuous functions on $\mathbb{R}^N$ vanishing at infinity.

Proposition 3.9. *For $2(N+1) > \beta$ and a matrix a satisfying (3.9), $Y^{N,a}$ is a Feller process, i.e.*

i) for every $t > 0$ and every $f \in C_0(\mathbb{R}^N)$ we have $Q_t f \in C_0(\mathbb{R}^N)$,

ii) for every $f \in C_0(\mathbb{R}^N)$, $\lim_{t \downarrow 0} Q_t f = f$ pointwise in $\mathbb{R}^N$.

Moreover, $Q_t f \in C(\mathbb{R}^N)$ for every $t > 0$ and every $f \in L^2(\mathbb{R}^N, \hat{q})$.

Remark 3.10. It is well known that i) and ii) even imply that $\lim_{t \downarrow 0} \|Q_t f - f\|_\infty = 0$ for each $f \in C_0(\mathbb{R}^N)$. Moreover, we can derive the following version of the strong Markov property (see Theorem 3 in Section 2.3 in [18]). Let T be a stopping time with $T \leq t_0$ a.s. for some $t_0 > 0$. Then, we have for each $f \in L^2(\mathbb{R}^N, \hat{q})$

$$E\left[f(Y_{t_0}) | \mathcal{F}_T\right] = E_{Y_T}\left[f(Y_{t_0 - T})\right],$$

with $(\mathcal{F}_t)_{t \geq 0}$ denoting the natural filtration of Y.

Proof of Proposition 3.9. ii) follows directly by path-continuity and dominated convergence. i) as well as the additional statement follow from the analytic regularity theory of symmetric diffusions, see [59], in particular Theorem 3.5, Proposition 3.1 and Corollary 4.2, provided the following two conditions are fulfilled:

- The measure $\hat{q}$ is doubling, i.e. there exists a constant C', such that for all Euclidian balls $B_R \subset B_{2R}$

$$\hat{q}(B_{2R}) \leq C' \hat{q}(B_R).$$

- $\hat{\mathcal{E}}$ satisfies a uniform local Poincaré inequality, i.e. there is a constant $C' > 0$ such that

$$\int_{B_R} |f - (f)_{B_R}|^2 d\hat{q} \leq C' R^2 \int_{B_R} |\nabla f|^2 d\hat{q},$$

for all Euclidian balls B_R and $f \in \mathcal{D}(\hat{\mathcal{E}})$, where $(f)_{B_R}$ denotes the integral $\frac{1}{\hat{q}(B_R)} \int_{B_R} f d\hat{q}$.

Both conditions are verified once we have proven that the weight function $\hat{q}$ is contained in the Muckenhoupt class $\mathcal{A}_2$, which will be done in Lemma 3.11 below. Indeed, the doubling property follows immediately from the Muckenhoupt condition (see e.g. [63] or [61]) and for the proof of the Poincaré inequality see Theorem 1.5 in [35]. □

Lemma 3.11. *For N such that $\beta < 2(N+1)$, we have $\hat{q} \in \mathcal{A}_2$, i.e. there exists a positive constant $C = C(N, \delta)$ such that for every Euclidian ball B_R*

$$\frac{1}{|B_R|} \int_{B_R} \hat{q} \, dx \, \frac{1}{|B_R|} \int_{B_R} \hat{q}^{-1} \, dx \leq C,$$

where $|B_R|$ denotes the Lebesgue measure of the ball B_R.

It suffices to prove the Muckenhoupt condition for the weight function $\tilde{q}$ defined by

$$\tilde{q}(x) := \prod_{i=1}^{N} (x^i - x^{i-1})^{\beta/(N+1)-1}, \quad x \in S_1, \tag{3.10}$$

and $\tilde{q}(x) := \tilde{q}(Tx)$ if $x \in \mathbb{R}^N$, since there exist positive constants C_1 and C_2 depending on δ and N such that

$$C_1 \tilde{q}(x) \leq \hat{q}(x) \leq C_2 \tilde{q}(x), \quad \forall x \in \mathbb{R}^N. \tag{3.11}$$

Note that $(1 - x_N)^{\frac{\beta}{N+1}-1}$ is uniformly bounded and bounded away from zero on Ω_N.

In the following we denote by $P_R(m)$ the parallelepiped in $\mathbb{R}^N$ with basis point $m = (m_1, \ldots, m_N)$, which is spanned by the vectors $v_i = \sum_{j=i}^{N} e_j$, $i = 1, \ldots, N$, normalized to the length R, where $(e_j)_{j=1,\ldots,N}$ is the canonical basis in $\mathbb{R}^N$. Then, $P_R(m)$ can also be written as

$$P_R(m) = m + \left\{ x \in \mathbb{R}^N : x^1 \in \left[0, \frac{R}{\sqrt{N}}\right], x^2 \in \left[x^1, x^1 + \frac{R}{\sqrt{N-1}}\right], \ldots, x^N \in \left[x^{N-1}, x^{N-1} + R\right] \right\}.$$

In order to prove the Muckenhoupt condition for $\tilde{q}$ we need the following

Lemma 3.12. *Let B_R be an arbitrary Euclidian ball in $\mathbb{R}^N$. Then, there exists a positive constant C only depending on N and a parallelepiped $P_{lR}(m) \subset S_1$ with $l > 0$ independent of B_R such that*

$$\int_{B_R} \tilde{q}^{\pm 1} dx \leq C \int_{P_{lR}(m)} \tilde{q}^{\pm 1} dx.$$

Proof. We denote by (S_n), $n = 1, \ldots, 2^N \cdot N!$, the subsets of $\mathbb{R}^N$ taking the form

$$S_n = \{x \in \mathbb{R}^N : (s_1 \pi_1(x), \ldots, s_N \pi_N(x)) \in S_1\},$$

where $s_i \in \{-1, 1\}$, $i = 1, \ldots, N$, and $\pi(x) = (\pi_1(x), \ldots, \pi_N(x))$ is a permutation of the components of x. Note that $\mathbb{R}^N = \bigcup_n S_n$ and the intersections of the sets S_n have zero Lebesgue measure.

Let now $B_R(M)$ be an arbitrary Euclidian ball with radius R centered in $M \in \mathbb{R}^N$. We first consider the case $|M| \leq 2R$. Then, obviously $B_R(M) \subset B_{4R}(0)$ and $B_{4R}(0) \cap S_1 \subset P_{4R}(0)$. Hence,

$$\int_{B_R(M)} \tilde{q}^{\pm 1} dx \leq \int_{B_{4R}(0)} \tilde{q}^{\pm 1} dx = \sum_n \int_{B_{4R}(0) \cap S_n} \tilde{q}^{\pm 1} dx.$$

By the definition of $\tilde{q}$ we have

$$\int_{B_{4R}(0) \cap S_n} \tilde{q}^{\pm 1} dx = \int_{B_{4R}(0) \cap S_1} \tilde{q}^{\pm 1} dx$$

for all $n \in \{1, \ldots, 2^N \cdot N!\}$. Thus,

$$\int_{B_R(M)} \tilde{q}^{\pm 1} dx \leq 2^N \cdot N! \int_{B_{4R}(0) \cap S_1} \tilde{q}^{\pm 1} dx \leq C \int_{P_{4R}(0)} \tilde{q}^{\pm 1} dx.$$

Suppose now that $|M| > 2R$. Let n_0 be such that $M \in S_{n_0}$. Then, by construction of $\tilde{q}$ we have

$$\int_{B_R(M) \cap S_n} \tilde{q}^{\pm 1} dx \leq \int_{B_R(M) \cap S_{n_0}} \tilde{q}^{\pm 1} dx$$

for all $n = 1, \ldots, 2^N \cdot N!$. Hence,

$$\int_{B_R(M)} \tilde{q}^{\pm 1} dx = \sum_n \int_{B_R(M) \cap S_n} \tilde{q}^{\pm 1} dx \leq 2^N \cdot N! \int_{B_R(M) \cap S_{n_0}} \tilde{q}^{\pm 1} dx.$$

Set $K := T(B_R(M) \cap S_{n_0}) \subset S_1$ with T defined as above. Then, it is clear by definition of $\tilde{q}$ that

$$\int_{B_R(M) \cap S_{n_0}} \tilde{q}^{\pm 1} dx = \int_K \tilde{q}^{\pm 1} dx$$

and we get

$$\int_{B_R(M)} \tilde{q}^{\pm 1} dx \leq C \int_K \tilde{q}^{\pm 1} dx.$$

Finally, we choose a parallelepiped $P_{2R}(m)$ such that $K \subseteq P_{2R}(m) \subset S_1$, which completes the proof. $\square$

Proof of Lemma 3.11. We prove the Muckenhoupt condition for $\tilde{q}$. Recall that we have assumed $\beta < 2(N+1)$. In the following the symbol C denotes a positive constant depending on N and δ with possibly changing its value from one occurence to another.

Using Lemma 3.12 we have

$$\frac{1}{|B_R|^2}\int_{B_R}\tilde{q}\,dx\int_{B_R}\tilde{q}^{-1}dx \leq CR^{-2N}\int_{P_{lR}(m)}\tilde{q}\,dx\int_{P_{lR}(m)}\tilde{q}^{-1}dx$$

$$=CR^{-2N}\int_{P_{lR}(m)}\prod_{i=1}^{N}(x^i-x^{i-1})^{\beta/(N+1)-1}dx$$

$$\times \int_{P_{lR}(m)}\prod_{i=1}^{N}(x^i-x^{i-1})^{-(\beta/(N+1)-1)}dx.$$

By the change of variables $y_i = x^i - x^{i-1}$, $i = 1,\ldots,N$, we obtain

$$\frac{1}{|B_R|^2}\int_{B_R}\tilde{q}\,dx\int_{B_R}\tilde{q}^{-1}dx \leq CR^{-2N}\prod_{i=1}^{N}\int_{\tilde{m}_i}^{\tilde{n}_i}y_i^{\frac{\beta}{N+1}-1}dy_i\int_{\tilde{m}_i}^{\tilde{n}_i}y_i^{-(\frac{\beta}{N+1}-1)}dy_i,$$

where we have set $\tilde{m}_i := m_i - m_{i-1}$ with $m_0 := 0$ and $\tilde{n}_i := \tilde{m}_i + \frac{lR}{\sqrt{N+1-i}}$ for abbreviation. Recall that in one dimension the weight function $x \mapsto |x|^\eta$ on $\mathbb{R}$ is contained in $\mathcal{A}_2$ if $\eta \in (-1,1)$ (see p. 229 and p. 236 in [61]). Hence, we get for every $i \in \{1,\ldots,N\}$

$$\int_{\tilde{m}_i}^{\tilde{n}_i}y_i^{\frac{\beta}{N+1}-1}dy_i \int_{\tilde{m}_i}^{\tilde{n}_i}y_i^{-(\frac{\beta}{N+1}-1)}dy_i \leq C\left|\frac{lR}{\sqrt{N+1-i}}\right|^2,$$

and the result follows.

□

3.3.2 Feller Properties of Y inside a Box

Let $E := E^\delta := \{x \in \mathbb{R}^N : \|x\|_\infty < 1 - 2\delta\}$ be a box in $\mathbb{R}^N$ centered in the origin and $\mathcal{B}(E)$ be the Borel σ-field on E. We denote by

$$\tau_E := \inf\{t > 0 : Y_t \in E^c\}$$

the first hitting time of E^c. This subsection is devoted to the proof of the Feller properties for the stopped process $Y_t^E := Y_t^{N,a,E} := Y_{t\wedge\tau_E}^{N,a}$, whose transition semigroup is given by

$$Q_t^E f(x) = E^x[f(Y_{t\wedge\tau_E})] = E^x\left[f(Y_t)\mathbb{1}_{\{t<\tau_E\}} + f(Y_{\tau_E})\mathbb{1}_{\{t\geq\tau_E\}}\right], \quad t > 0, \, x \in \bar{E},$$

for every bounded f on $\bar{E}$. In order to prove the Feller property we will follow essentially the proof of Theorem 13.5 in [30] (cf. also [19]). It is shown there that the Feller properties are preserved, if the domain is regular in the following sense.

Proposition 3.13. *The domain E is regular, i.e. for every $z \in \partial E$ we have $P^z[\tau_E = 0] = 1$.*

Remark 3.14. $z \in \partial E$ is regular point in the sense of the definition given in Proposition 3.13 if and only if for every continuous function f on ∂E

$$\lim_{E\ni x\to z}E^x[f(Y_{\tau_E})\mathbb{1}_{\{\tau_E<\infty\}}] = f(z).$$

For the Brownian case we refer to Theorem 2.12 in [45] and to Theorem 1.23 in [21]. The arguments are robust, but it is required that under P^x the probability of the event that the exit time of a ball centered at x does not exceed t is arbitrary small uniformly in x for a suitable chosen small $t > 0$. In our situation this property is ensured by [58], p. 330.

The proof of this proposition will be based on the following Wiener test established by Fabes, Jerison and Kenig.

Theorem 3.15. *Let $B_{R_0}(0)$ be a large ball centered in zero such that $E \subset B_{R_0/4}(0)$. Then, a point $z \in \partial E$ is regular if and only if*

a) $\int_0^{R_0} \frac{s^2}{\hat{q}(B_s(z))} \frac{ds}{s} < \infty$, *or*

b) $\int_0^{R_0} \mathrm{cap}(K_\rho) \frac{\rho^2}{\hat{q}(B_\rho(z))} \frac{d\rho}{\rho} = \infty$,

where $K_\rho(z) := (B_{R_0}(0) \setminus E) \cap B_\rho(z)$ and cap *denotes the capacity associated with the Dirichlet form $\mathcal{E}^{N,a}$.*

Proof. See Theorem 5.1 in [34]. □

In the sequel we will use the following notation

$$d(x) := \left| \left\{ i \in \{1, \ldots, N\} : x^i = x^{i-1} \right\} \right|, \quad x \in \mathbb{R}^N,$$

again with the convention $x^0 := 0$. In order to prove the regularity of E we start with a preparing lemma.

Lemma 3.16. *Let $y \in \mathbb{R}^N$ be arbitrary.*

i) There exists a positive $r_0 = r_0(y)$ such that for all balls $B_r(y)$, $0 < r \leq r_0$, there exists a parallelepiped $P_{lr}(\bar{y})$ contained in S_1 with $d(\bar{y}) = d(y)$, a positive constant C and $l > 0$, such that

$$\hat{q}(B_r(y)) \leq C \hat{q}(P_{lr}(\bar{y})).$$

ii) For all balls $B_r(y)$ there exists parallelepiped $P_{lr}(\bar{y})$ contained in S_1 with $d(\bar{y}) = d(y)$, for some $l > 0$ such that $\hat{q}(P_{lr}(\bar{y})) \leq \hat{q}(B_r(y))$.

iii) For all y we have $C_1 r^{h(y)} \leq \hat{q}(B_r(y))$ for all r and $\hat{q}(B_r(y)) \leq C_2 r^{h(y)}$ for all $r \leq r_0(y)$ for some positive constants C_1 and C_2 depending on y, where

$$h(y) := N - d(y) + \tfrac{\beta}{N+1} d(y).$$

Proof. Obviously, due to (3.11) it suffices to prove the lemma with $\hat{q}$ replaced by $\tilde{q}$ defined in (3.10).

i) The case $y = 0$ is clear. For $y \neq 0$ we choose r_0 such that $2r_0 \leq |y|$ and let n_0 be such that $y \in S_{n_0}$. Consider now an arbitrary ball $B_r(y)$ with $r \leq r_0$ and let $K = T(S_{n_0} \cap B_r(y))$ be the subset of S_1 constructed in the second part of the proof of Lemma 3.12. Then, possibly after choosing a smaller r_0 we find a parallelepiped and a positive constant l such that $K \subset P_{lr}(\bar{y}) \subset S_1$ with $d(\bar{y}) = d(y)$ and we obtain i).

ii) Let $K = T(S_{n_0} \cap B_r(y))$ be defined as in i). Then, clearly $\tilde{q}(B_r(y)) \geq \tilde{q}(K)$. Thus, we can choose $\bar{y} = Ty$ and l independent of r such that $P_{lr}(\bar{y}) \subseteq K$ and ii) follows.

iii) We proceed similar as in the proof of Lemma 3.11. For some parallelepiped $P_{lr}(\bar{y})$ with $d(\bar{y}) = d(y)$ we use a change of variables to obtain

$$\tilde{q}(P_{lr}(\bar{y})) = \int_{P_{lr}(\bar{y})} \prod_{i=1}^{N} (x^i - x^{i-1})^{\frac{\beta}{N+1}-1} dx = \prod_{i=1}^{N} \int_{\tilde{y}_i}^{\tilde{y}_i + c_i r} z_i^{\frac{\beta}{N+1}-1} dz_i,$$

where $\tilde{y}_i = \bar{y}_i - \bar{y}_{i-1}$, $\bar{y}_0 := 0$ and $c_i := \frac{l}{\sqrt{N+1-i}}$. Note that $d(\bar{y}) = d(y)$ is the number of components of $\tilde{y}$ which are equal to zero. Using the mean value theorem we obtain that

$$C_1 r^{h(y)} \leq \hat{q}(P_{lr}(\bar{y})) \leq C_2 r^{h(y)}$$

for some positive constants C_1 and C_2 depending on y, so that iii) follows from i) and ii). □

Proof of Proposition 3.13. Let $z \in \partial E$ be fixed and R_0 be as in the statement of Theorem 3.15. Setting $h'(z) := 1 - h(z)$ let us first consider the case $h'(z) > -1$. Then, using Lemma 3.16 iii) we have that

$$\int_0^{R_0} \frac{s^2}{\hat{q}(B_s(z))} \frac{ds}{s} \leq C \int_0^{r_0} s^{h'(z)} ds + \frac{1}{2\hat{q}(B_{r_0}(z))}(R_0^2 - r_0^2) < \infty,$$

with $r_0 = r_0(z)$ as above in Lemma 3.16 iii). Thus, the criterion a) in Theorem 3.15 applies and the regularity of z follows. The case $h'(z) \leq -1$ is more difficult. Combining Lemma 3.1 in [35] and Theorem 3.3 in [35] and using Lemma 3.16 iii) we get the following estimate for the capacity of small balls:

$$\text{cap}(B_r(y)) \simeq \left(\int_r^{R_0} \frac{s^2}{\hat{q}(B_s(y))} \frac{ds}{s} \right)^{-1} \geq C \left(\int_r^{R_0} s^{h'(y)} ds \right)^{-1}. \tag{3.12}$$

Recall the definition of the set $K_\rho(z)$ in Theorem 3.15. Clearly, for every ρ sufficiently small there exists a ball $B_{\rho/2}(\hat{z})$ with $\hat{z}$ depending on ρ and with $d(\hat{z}) = d(z)$ such that $B_{\rho/2}(\hat{z}) \subset K_\rho(z)$. Let now r_0 be such that Lemma 3.16 iii) and (3.12) hold for every ball $B_\rho(z)$, $\rho \leq r_0$. Then, we obtain in the case $h'(z) < -1$

$$\int_0^{R_0} \text{cap}(K_\rho) \frac{\rho^2}{\hat{q}(B_\rho(z))} \frac{d\rho}{\rho} \geq C \int_0^{r_0} \text{cap}(B_{\rho/2}(\hat{z})) \rho^{h'(z)} d\rho \geq C \int_0^{r_0} \left(\int_{\rho/2}^{R_0} s^{h'(\hat{z})} ds \right)^{-1} \rho^{h'(z)} d\rho$$

$$= C \left(h'(z) + 1 \right) \int_0^{r_0/2} \frac{\rho^{h'(z)}}{R_0^{h'(y)+1} - \rho^{h'(y)+1}} d\rho$$

$$= C \int_0^{r_0/2} \left(-\log(|R_0^{h'(y)+1} - \rho^{h'(y)+1}|) \right)' d\rho = \infty.$$

Finally, if $h'(z) = -1$ we get by an analogous procedure

$$\int_0^{R_0} \operatorname{cap}(K_\rho) \frac{\rho^2}{\hat{q}(B_\rho(z))} \frac{d\rho}{\rho} \geq C \int_0^{r_0/2} \frac{1}{\rho(\log R_0 - \log \rho)} d\rho$$
$$= C \int_0^{r_0/2} \left(-\log(\log R_0 - \log \rho) \right)' d\rho = \infty.$$

Hence, applying the criterion b) of Theorem 3.15 completes the proof. □

Proposition 3.17. Y^E *is a Feller process, i.e.*

i) for every $t > 0$ and every $f \in C(\bar{E})$ we have $Q_t^E f \in C(\bar{E})$,

ii) for every $f \in C(\bar{E})$, $\lim_{t \downarrow 0} Q_t^E f = f$ pointwise in $\bar{E}$.

The statement is classical and can be found e.g. in Theorem 13.5 in [30]. For illustration we repeat the argument here. We shall need the following lemma (cf. [18], p. 73, Exercise 2).

Lemma 3.18. *For any compact set $K \subset E$ we have*

$$\lim_{t \downarrow 0} \sup_{x \in K} P^x[\tau_E \leq t] = 0.$$

Proof. We need to show that for any $\delta > 0$ there exists a $t_0 > 0$ such that

$$\inf_{x \in K} P^x[\tau_E \geq t_0] \geq 1 - \delta. \tag{3.13}$$

Consider a bounded function $f \in C_0(\mathbb{R}^N)$ such that $0 \leq f \leq 1$, $f = 1$ on K and $f = 0$ on the complement of E. Let now t_0 be such that $\sup_{t \leq t_0} \|Q_t f - f\|_\infty < \delta/2$ (cf. Remark 3.10). Then,

$$E^x[f(Y_{t_0})] \leq P^x[\tau_E \geq t_0] + E^x[f(Y_{t_0}) \mathbb{1}_{\{\tau_E < t_0\}}].$$

For $x \in K$ the left hand side is equal to

$$Q_{t_0} f(x) = 1 + Q_{t_0} f(x) - f(x) \geq 1 - \sup_{t \leq t_0} \|Q_t f - f\|_\infty \geq 1 - \frac{\delta}{2}.$$

On the other hand, using the strong Markov property (cf. again Remark 3.10) and the fact that $f(Y_{\tau_E}^N) = 0$ we have

$$E^x[f(Y_{t_0}) \mathbb{1}_{\{\tau_E < t_0\}}] = E^x\left[E^x[f(Y_{t_0})|\mathcal{F}_{\tau_E}] \mathbb{1}_{\{\tau_E < t_0\}}\right] = E^x\left[Q_{t_0 - \tau_E} f(Y_{\tau_E}) \mathbb{1}_{\{\tau_E < t_0\}}\right]$$
$$= E^x\left[(Q_{t_0 - \tau_E} f(Y_{\tau_E}) - f(Y_{\tau_E})) \mathbb{1}_{\{\tau_E < t_0\}}\right] \leq E^x\left[\sup_{t \leq t_0} \|Q_t f - f\|_\infty \mathbb{1}_{\{\tau_E < t_0\}}\right]$$
$$\leq \frac{\delta}{2},$$

and (3.13) follows. □

Proof of Proposition 3.17. Let $t > 0$ and $f \in C(\bar{E})$. Then, by the semigroup property of (Q_t^E) we have for $0 < s < t$
$$Q_t^E f(x) = E^x \left[\psi_s(Y_{s \wedge \tau_E}) \right],$$
where
$$\psi_s(x) = Q_{t-s}^E f(x) = E^x \left[f(Y_{(t-s) \wedge \tau_E}) \right], \quad x \in \bar{E}.$$
Then, ψ_s can be extended to a function in $L^2(\mathbb{R}^N, \hat{q})$ and by Proposition 3.9 we have $Q_s \psi_s \in C_b(\mathbb{R}^N)$. Since
$$\left| Q_t^E f(x) - Q_s \psi_s(x) \right| = \left| E^x[\psi_s(Y_{s \wedge \tau_E}) - \psi_s(Y_s)] \right| \le 2 \left\| \psi_s \right\| P^x[\tau_E \le s] \le 2 \left\| f \right\| P^x[\tau_E \le s]$$
and since the right hand side converges to zero uniformly in x on every compact subset of E by Lemma 3.18, we conclude that $Q_t^E f \in C_b(E)$, i.e. $Q_t^E f$ is continuous in the interior of E. In order to show i) it suffices to verify continuity at the boundary. Since E is regular, we have obviously $Q_t^E f = f$ on ∂E. By Lemma 13.1 in [30] we have upper semicontinuity of the mapping $x \mapsto P^x[t < \tau_E]$. Hence, we obtain for $z \in \partial E$,
$$\limsup_{x \to z} P^x[t < \tau_E] \le P^z[t < \tau_E] = 0,$$
where we have used the regularity of E in Proposition 3.13. Thus, for every $x \in E$
$$\left| Q_t^E f(x) - f(z) \right| \le \left| Q_t^E f(x) - E^x[f(Y_{\tau_E}) \mathbb{1}_{\{\tau_E < \infty\}}] \right| + \left| E^x[f(Y_{\tau_E}) \mathbb{1}_{\{\tau_E < \infty\}}] - f(z) \right|$$
$$\le \left| E^x \left[f(Y_t) \mathbb{1}_{\{t < \tau_E\}} + f(Y_{\tau_E}) \mathbb{1}_{\{t \ge \tau_E\}} - f(Y_{\tau_E}) \mathbb{1}_{\{\tau_E < \infty\}} \right] \right|$$
$$+ \left| E^x[f(Y_{\tau_E}) \mathbb{1}_{\{\tau_E < \infty\}}] - f(z) \right|$$
$$\le \|f\|_\infty P^x[t < \tau_E] + \left| E^x \left[f(Y_{\tau_E}) \mathbb{1}_{\{\tau_E < \infty\}} \left(\mathbb{1}_{\{t \ge \tau_E\}} - 1 \right) \right] \right|$$
$$\le 2 \|f\|_\infty P^x[t < \tau_E] + \left| E^x[f(Y_{\tau_E}) \mathbb{1}_{\{\tau_E < \infty\}}] - f(z) \right|,$$
where we have used the fact that the event $\{\tau_E < \infty\}$ is included in $\{t \ge \tau_E\}$. Since z is a regular point, the second term tends to zero as $x \to z$, $x \in E$, cf. Remark 3.14. Hence,
$$\lim_{E \ni x \to z} Q_t^E f(x) = f(z)$$
and property i) is proven.

To prove ii), we extend f to a function in $C_0(\mathbb{R}^N)$, i.e. we may deduce from Proposition 3.9 that $\lim_{t \downarrow 0} Q_t f(x) = f(x)$ for every $x \in E$. Furthermore, we have for every $x \in E$
$$|Q_t^E f(x) - Q_t f(x)| \le P^x[t \le \tau_E] \|f\|_\infty,$$
and $P^x[\tau_E = 0] = 0$, since E is open and Y has continuous paths. Hence,
$$\lim_{t \downarrow 0} Q_t^E f(x) = \lim_{t \downarrow 0} Q_t f(x) = f(x).$$
This gives pointwise convergence on E. Since $Q_t^E f = f$ on ∂E by regularity, the convergence on ∂E is trivial. □

3.3.3 Feller Properties of X^N

In this section we will finally prove the Feller property for X^N stated in Theorem 3.2. To this end we construct a Feller process $\tilde{X}$ taking values in $\overline{\Sigma}_N$ and in a second step we will identify this process with X^N.

In analogy to the definition of Ω_N above we set

$$\Omega_i := \Omega_i^\delta := \{x \in \overline{\Sigma}_N : x^{i+1} - x^i \geq \delta\}, \qquad i = 0, \ldots, N. \tag{3.14}$$

and moreover

$$A^i := \partial \Omega_i^{2\delta} \setminus \partial \Sigma_N = \{x \in \overline{\Sigma}_N : x^{i+1} - x^i = 2\delta\}.$$

Notice that we can choose δ so small that $\overline{\Sigma}_N = \bigcup_{i=1}^N \Omega_i^{2\delta}$. Furthermore, we define the mappings H_i, $i = 0, \ldots, N$, by

$$H_i(x) := \left(x^1, \ldots, x^i, 1 - (x^N - x^i), 1 - (x^{N-1} - x^i), \ldots, 1 - (x^{i+1} - x^i)\right), \qquad x \in \overline{\Sigma}_N.$$

Notice that for every i, H_i maps Ω_N on Ω_i and vice versa. In particular, $H_i \circ H_i = \text{Id}$ and H_N is the identity on Ω_N. Let $T : \bar{E} \to \Omega_N^{2\delta} \subset \Omega_N$ be defined as in (3.7). Let Y^i denote the $\mathbb{R}^N$-valued Feller process induced from the form (3.8) with $a = a_i$, where the matrix valued function a_i is defined $\hat{q}$-almost everywhere by

$$a_i(x) := \begin{cases} DH_i^t & \text{for } x \in \mathring{S}_1 \\ DH_i^t \cdot (DT_{|S_n}^{-1})^t & \text{for } x \in \mathring{S}_n,\, n \in \{2, \cdots 2^N \cdot N!\}, \end{cases} \tag{3.15}$$

such that condition (3.9) is clearly satisfied (note that $DH_i = DH_i^{-1}$). In particular, a_i is constant on every S_n. Moreover, setting $\rho_i := H_i \circ T$, clearly $a_i = (D\rho_i^{-1})^t$.

Remark 3.19. Analogously to Proposition 3.7 one can establish the following integration by parts formula associated to $\mathcal{E}^{N,a_i}$. Let $u \in C^1(\mathbb{R}^N)$ and ξ a continuous vector field on $\mathbb{R}^N$ such that $\text{supp}\,\xi \subseteq \{x \in \mathbb{R}^N : \|x\|_\infty < 1 - \delta\}$, ξ is continuously differentiable in the interior of each S_n and ξ satisfies the boundary condition $\langle a_i^t \cdot a_i \cdot \xi, \nu_n \rangle = 0$ on ∂S_n for every n, ν_n denoting the outward normal field of ∂S_n. Then,

$$\int_{\mathbb{R}^N} \langle a_i \nabla u, a_i \xi \rangle \hat{q}(dx) = -\int_{\mathbb{R}^N} u \left[\text{div}(a_i^t \cdot a_i \cdot \xi) + \langle a_i^t \cdot a_i \cdot \xi, \nabla \log \hat{q} \rangle \right] \hat{q}(dx).$$

Thus, every smooth function g such that $\xi = \nabla g$ satisfies the above conditions is contained in the domain of the generator L^i associated to $\mathcal{E}^{N,a_i}$ and on $\mathbb{R}^N \setminus \bigcup_n \partial S_n$ we have

$$L^i g = \text{div}(a_i^t \cdot a_i \cdot \nabla g) + \langle a_i^t \cdot a_i \cdot \nabla g, \nabla \log \hat{q} \rangle.$$

Next we define the $\Omega_i^{2\delta}$-valued process $\tilde{X}^i$ by

$$\tilde{X}_t^i = \rho_i(Y_t^{i,E}) = H_i \circ T(Y_t^{i,E}), \qquad t \geq 0.$$

The semigroup of $\tilde{X}^i$ will be denoted by $(\tilde{P}_t^i)_{t \geq 0}$, i.e. $\tilde{P}_t^i f(x) = E^x[f(\tilde{X}_t^i)]$.

Lemma 3.20. *For every i, $\tilde{X}^i$ is Markovian.*

Proof. Since H_i is an injective mapping for every i, it suffices to show that the process $T(Y^{i,E})$ is Markovian. Moreover, since $T(Y_t^{i,E}) = T(Y_{t\wedge\tau_E}^i) = (T(Y_\cdot^i))_{t\wedge\tau_E}$ it is enough to prove the Markov property for the process $T(Y_\cdot^i)$, which is implied e.g. by the condition that for any Borel set $A \subseteq \overline{\Sigma}_N$

$$P^x[Y_t^i \in T^{-1}(A)] = P^y[Y_t^i \in T^{-1}(A)] \quad \text{whenever } T(x) = T(y). \tag{3.16}$$

Now the choice of $\hat{q}$ and the metric a_i imply for any Borel set $A \subseteq \overline{\Sigma}_N$ condition (3.16) is satisfied. To see this, let $\{\sigma_k \mid k = 1, \cdots, N + N(N-1)/2\}$ be the collection of line-reflections in $\mathbb{R}^N$ with respect to either one of the coordinate axes $\{\lambda e_i, \lambda \in \mathbb{R}\}$ or a diagonal $\{\lambda(e_j + e_k)\}$, then for $x, y \in \mathbb{R}^N$ with $T(x) = T(y)$ there exists a finite sequence $\sigma_{k_1}, \cdots \sigma_{k_l}$ such that $\tau(x) := \sigma_{k_1} \circ \sigma_{k_2} \cdots \circ \sigma_{k_l}(x) = y$. Now each of the reflections σ_i preserves the Dirichlet form (3.8) when a is chosen as in (3.15), such that the processes $\tau(Y^{i,x})$ and $Y^{i,y}$ are equal in distribution. Moreover, $\tau(T^{-1}(A)) = T^{-1}(A)$, from which (3.16) is obtained. $\square$

Lemma 3.21. *For each $f \in C_{Neu}^2$ the process $t \to f(\tilde{X}_t^i) - \int_0^t L^N f(\tilde{X}_s^i)\,ds$ is a martingale w.r.t. the filtration generated by $\tilde{X}^i$.*

Proof. Similar to Proposition 3.7 one checks for $f \in C_{Neu}^2 \cap C_c(\Omega_i)$ that the function f_i on $\mathbb{R}^N$, which is defined by $f_i = f \circ H_i \circ T = f \circ \rho_i$ on the set $\{x \in \mathbb{R}^N : \|x\|_\infty < 1 - \delta\}$ and $f_i = 0$ on the complement this set, belongs to the domain of the generator L^i of the Dirichlet form (3.8) with $a = a_i$ as in (3.15) (cf. Remark 3.19). Hence the process $f_i(Y^i) - \int_0^\cdot L^i f_i(Y_s^i)ds$ is a martingale w.r.t. to the filtration generated by Y^i and thus also $f_i(Y^{i,E}) - \int_0^\cdot L^i f_i(Y_s^{i,E})ds$ due to the optional sampling theorem. Obviously in the last statement the function f can be modified outside of $\Omega_i^{2\delta}$, i.e. it holds also for $f \in C_{Neu}^2$. Moreover, $f_i(Y^{i,E}) = f(\tilde{X}_\cdot^i)$ and $L^i f_i = (L^N f) \circ \rho_i$ on $\bar{E}$. Indeed, since $a_i = (D\rho_i^{-1})^t$, $\nabla(f \circ \rho_i) = (D\rho_i)^t \cdot \nabla f \circ \rho_i$ and $\hat{q}(x) = q_N(T(x)) = q_N(\rho_i(x))$ for all $x \in \bar{E}$, we obtain

$$L^i f_i = \operatorname{div}(a_i^t \cdot a_i \cdot (D\rho_i)^t \cdot \nabla f \circ \rho_i) + \langle a_i^t \cdot a_i \cdot (D\rho_i)^t \cdot \nabla f \circ \rho_i, \nabla(\log q_N \circ \rho_i)\rangle$$
$$= \operatorname{div}(D\rho_i^{-1} \cdot \nabla f \circ \rho_i) + \langle D\rho_i^{-1} \cdot \nabla f \circ \rho_i, D\rho_i^t \cdot \nabla \log q_N \circ \rho_i\rangle$$
$$= \Delta f \circ \rho_i + \langle \nabla f \circ \rho_i, \nabla \log q_N \circ \rho_i\rangle = (L^N f) \circ \rho_i.$$

Thus, $f(\tilde{X}_t^i) - \int_0^\cdot L^N f(\tilde{X}_s^i)ds$ is a martingale w.r.t. the filtration generated by Y^i and adapted to the filtration generated by $\tilde{X}_\cdot^i = H_i \circ T(Y^{i,E})$ which establishes the claim. $\square$

Proposition 3.22. *For every i, $\tilde{X}^i$ is a Feller process, more precisely*

i) for every $t > 0$ and every $f \in C(\Omega_i^{2\delta})$ we have $\tilde{P}_t^i f \in C(\Omega_i^{2\delta})$,

ii) for every $f \in C(\Omega_i^{2\delta})$, $\lim_{t\downarrow 0} \tilde{P}_t^i f = f$ pointwise in $\Omega_i^{2\delta}$.

Proof. Since obviously for every continuous f on $\Omega_i^{2\delta}$

$$\tilde{P}_t^i f(x) = E^x[f(\tilde{X}_t^i)] = E^x[f(H_i \circ T(Y_t^{i,E}))] = Q_t^E(f \circ H_i \circ T)(x), \quad t > 0,$$

the result follows from Proposition 3.17 and the continuity of the mappings H_i and T. $\square$

3.3 Feller Property

Next we define the process $\tilde{X}$ with state space $\overline{\Sigma}_N$ as follows: Let q_N be the initial distribution of $\tilde{X}$, i.e. $\tilde{X}_0 \sim q_N$. Choose $i_1 \in \{0, \ldots N\}$ such that $\tilde{X}_0 \in \Omega_{i_1}^{2\delta}$ and $\text{dist}(\tilde{X}_0, A^{i_1}) = \max_i \text{dist}(\tilde{X}_0, A^i)$. We set $\tilde{X}_t = \tilde{X}_t^{i_1}$ for $0 \leq t \leq T_1$, where T_1 denotes the first hitting time of A^{i_1}, i.e. on $[0, T_1]$ the process behaves according to $\tilde{P}^{i_1}$. Choose now $i_2 \in \{0, \ldots N\}$ such that $\tilde{X}_{T_1} \in \Omega_{i_2}^{2\delta}$ and $\text{dist}(\tilde{X}_{T_1}, A^{i_2}) = \max_i \text{dist}(\tilde{X}_{T_1}, A^i)$. At time T_1 the process starts afresh from $\tilde{X}_{T_1}$ according to $(\tilde{P}_t^{i_2})$ up to the first time T_2 after T_1, when it hits A^{i_2}. This procedure is repeated forever.

Proposition 3.23. $\tilde{X}$ *is a Feller process.*

Proof. Let $(\tilde{P}_t)$ denote the semigroup associated to $\tilde{X}$. For $f \in C(\overline{\Sigma}_N)$, we need to show that $\tilde{P}_t f \in C(\overline{\Sigma}_N)$ for every $t > 0$. Let us first show that $\tilde{P}_{T_n} f \in C(\overline{\Sigma}_N)$ for every n using an induction argument. For an arbitrary $x \in \overline{\Sigma}_N$, choose i_1 as above depending on x such that $\tilde{P}_{T_1} f(x) = \tilde{P}_{T_1}^{i_1} f(x)$. Since $\tilde{P}_{T_1}^{i_1} f \in C(\Omega_{i_1}^{2\delta})$, we conclude that $\tilde{P}_{T_1} f$ is continuous in x for every x. For arbitrary n and $x \in \overline{\Sigma}_N$ we have by the strong Markov property

$$\tilde{P}_{T_{n+1}} f(x) = E^x[f(\tilde{X}_{T_{n+1}})] = E^x\left[E^{\tilde{X}_{T_N}}[f(\tilde{X}_{T_{n+1}-T_n}^{i_n})]\right] = E^x\left[\tilde{P}_{T_{n+1}-T_n}^{i_n} f(\tilde{X}_{T_N})\right]$$
$$= \tilde{P}_{T_n}(P_{T_{n+1}-T_n}^{i_n} f)(x)$$

and since $P_{T_{n+1}-T_n}^{i_n} f$ can be extended to a continuous function on $\overline{\Sigma}_N$, we get $\tilde{P}_{T_{n+1}} f \in C(\overline{\Sigma}_N)$ by the induction assumption.

Similarly, one can show that for every n the mapping $x \mapsto E^x[f(\tilde{X}_t)\mathbb{1}_{\{t \in (T_n, T_{n+1}]\}}]$ is continuous. Finally, for every $x \in \overline{\Sigma}_N$

$$\left|\tilde{P}_t f(x) - \sum_{k=0}^{n-1} E^x\left[f(\tilde{X}_t)\mathbb{1}_{\{t \in (T_k, T_{k+1}]\}}\right]\right| = \left|E^x\left[f(\tilde{X}_t)\mathbb{1}_{\{t > T_n\}}\right]\right| \leq \|f\|_\infty P^x[t > T_n]$$

and since $T_n \nearrow \infty$ P^x-a.s. locally uniformly in x as n tends to infinity, the claim follows. $\square$

The proof of Theorem 3.2 is complete, once we have shown that the processes $\tilde{X}$ and the original process X^N have the same law.

Lemma 3.24. *For $x \in \overline{\Sigma}_N$ the process $(\tilde{X}_{\cdot}^x)$ obtained from conditioning $\tilde{X}$ to start in x, solves the martingale problem for the operator (L^N, C_{Neu}^2) as in (3.2) and starting distribution δ_x.*

Proof. Let $\{T_k\}$ be the sequence of strictly increasing stopping times introduced in the construction of the process $\tilde{X}$. Then for $s < t$

$$f(\tilde{X}_t) - f(\tilde{X}_s) - \int_s^t L^N f(\tilde{X}_\sigma) d\sigma = \sum_k f(\tilde{X}_{(T_{k+1} \vee s) \wedge t}) - f(\tilde{X}_{(T_k \vee s) \wedge t}) - \int_{(T_k \vee s) \wedge t}^{(T_{k+1} \vee s) \wedge t} L^N f(\tilde{X}_\sigma) d\sigma.$$

Hence

$$\mathbb{E}(f(\tilde{X}_t) - \int_0^t L^N f(\tilde{X}_\sigma) d\sigma \,|\mathcal{F}_s) = f(\tilde{X}_s) - \int_0^s L^N f(\tilde{X}_\sigma) d\sigma$$

$$+ \sum_k \mathbb{E}\big(f(\tilde{X}_{(T_{k+1}\vee s)\wedge t}) - f(\tilde{X}_{(T_k\vee s)\wedge t}) - \int_{(T_k\vee s)\wedge t}^{(T_{k+1}\vee s)\wedge t} L^N f(\tilde{X}_\sigma) d\sigma \,|\mathcal{F}_s\big).$$

Using the strong Markov property of the Feller process $\tilde{X}$ and its pathwise decomposition into pieces of $\{\tilde{X}^i\}$-trajectories one obtains

$$\mathbb{E}\big(f(\tilde{X}_{(T_{k+1}\vee s)\wedge t}) - f(\tilde{X}_{(T_k\vee s)\wedge t}) - \int_{(T_k\vee s)\wedge t}^{(T_{k+1}\vee s)\wedge t} L^N f(\tilde{X}_\sigma) d\sigma \,|\mathcal{F}_s\big)$$

$$= \mathbb{E}\big[\mathbb{E}_{\tilde{X}_{(T_k\vee s)\wedge t}}\big(f(\tilde{X}^{i_k}_{\tau\wedge(t-s)}) - f(\tilde{X}^{i_k}_0) - \int_0^{\tau\wedge(t-s)} L^N f(\tilde{X}^{i_k}_\sigma) d\sigma\big) \,|\mathcal{F}_s\big], \quad (3.17)$$

where, by construction of $\tilde{X}$, τ ist the hitting time of the the set A^{i_k} for which $\text{dist}(A^{i_k}, \tilde{X}_{T_k}) = \max_i \text{dist}(A^i, \tilde{X}_{T_k})$. Lemma 3.21 implies that the inner expectation in (3.17) is zero. $\square$

The last ingredient for our proof of Theorem 3.2 is the identification of the processes $\tilde{X}$ with X.. Since both are Markovian and solve the martingale problem for the operator (L^N, C^2_{Neu}) it suffices to show that the martingale problem admits at most one Markovian solution. Clearly, any such solution induces a symmetric sub-Markovian semigroup on $L^2(\Sigma_N, q_N)$ whose generator extends (L^N, C^2_{Neu}). Hence it is enough to establish the following so-called Markov uniqueness property of (L^N, C^2_{Neu}), cf. [31, Definition 1.2], stated in Proposition 3.3.

Proof of Proposition 3.3. Let $H^{1,2}(\Sigma_N, q_N)$ (resp. '$H^{1,2}_0(\Sigma_N, q_N)$' in the notation of [31]) denote the closure of C^2_{Neu} w.r.t. to the norm $\|f\|_1 = (\|f\|^2_{L^2(\Sigma_N, q_N)} + \|\nabla f\|^2_{L^2(\Sigma_N, q_N)})^{1/2}$ and let $W^{1,2}$ be the Hilbert space of $L^2(\Sigma_N, q_N)$-functions f whose distributional derivative Df is in $L^2(\Sigma_N, q_N)$, equipped with the norm $\|f\|_1 = (\|f\|^2_{L^2(\Sigma_N, q_N)} + \|Df\|^2_{L^2(\Sigma_N, q_N)})^{1/2}$. Clearly, the quadratic form $Q(f, f) = \langle Df, Df \rangle_{L^2(\Sigma_N, q_N)}$, $\mathcal{D}(Q) = W^{1,2}(\Sigma_N, q_N)$, is a Dirichlet form on $L^2(\Sigma_N, q_N)$. Hence we may use the basic criterion for Markov uniqueness [31, Corollary 3.2], according to which (L, C^2_{Neu}) is Markov unique if $H^{1,2} = W^{1,2}$.

To prove the latter it obviously suffices to prove that $H^{1,2}$ is dense in $W^{1,2}$. Again we proceed by localization as follows. For fixed $\delta > 0$ let $\Omega^\delta_i = \Omega_i$ denote the subsets defined in (3.14), then $\{\Omega^{3\delta}_i\}$ constitutes a relatively open covering of $\overline{\Sigma}_N$ for δ small enough. Let $\{\eta_i\}_{i=0,\cdots,N}$ and $\{\chi_i\}_{i=0,\cdots,N}$ be collections of smooth cut-off function on $\overline{\Sigma}_N$ such that $\eta_i = 1$ on $\Omega^{3\delta}_i$ and $\text{supp}(\eta_i) \subset \Omega^{2\delta}_i$ and $\chi_i = 1$ on $\Omega^{2\delta}_i$ and $\text{supp}(\chi_i) \subset \Omega^\delta_i$ respectively.

Step 1. In the first step we reduce the problem to functions in $W^{1,2}$, whose support is contained in one of the sets $\Omega^{2\delta}_i$. For arbitrary $f \in W^{1,2}$ we have $f = f \cdot \eta_N + f \cdot (1 - \eta_N) =: f_N + r_N$. Next we write $r_N = r_N \cdot \eta_{N-1} + r_N \cdot (1 - \eta_{N-1}) =: f_{N-1} + r_{N-1}$. Iterating this procedure yield the decomposition $f = \sum_i f_i$ with $f_i \in W^{1,2}(\Sigma_N, q_N)$ and $\text{supp}(f_i) \subset \Omega^{2\delta}_i$. Hence, it suffices to prove that $f_i \in H^{1,2}(\Sigma_N, q_N)$. In Step 2 we will show first that f_i can be approximated w.r.t.

$\|.\|_1$ by functions which are smooth up to boundary and in Step 3 that such functions can be approximated in $\|.\|_1$ by smooth Neumann functions. We give details for the case $i = N$ only, the other cases can be treated almost the same way by using the maps H_i.

Step 2. Let $f_N \in W^{1,2}(\Sigma_N, q_N)$ with $\text{supp}(f_N) \subset \Omega_N^{2\delta}$. Then, obviously $f_N = f_N \cdot \chi_N$ and the restriction of f_N to Ω_N^{δ} belongs to the space $W^{1,2}(\Omega_N^{\delta}, q_N) = W^{1,2}(\Omega_N^{\delta}, \hat{q})$, where $\hat{q}$ denotes the modification of q_N according to (3.6). Due to Lemma 3.11 the extension $\hat{q}(x) = \hat{q}(T(x))$, $x \in \mathbb{R}^N$, lies in the Muckenhoupt class $\mathcal{A}_2$. Further note that $W^{1,2}(\Omega_N^{\delta}, \hat{q}) \subset W^{1,2}(\Omega_N^{\delta}, dx) = H^{1,2}(\Omega_N^{\delta}, dx)$ if $\beta \leq N+1$, and $W^{1,2}(\Omega_N^{\delta}, \hat{q}) \subset W^{1,1}(\Omega_N^{\delta}, dx) = H^{1,1}(\Omega_N^{\delta}, dx)$ by the Hölder inequality if $N+1 < \beta < 2(N+1)$, such that f_N has well defined boundary values in $L^1(\partial\Omega_N, dx)$. Hence we may conclude that the extension $\hat{f}_N(x) = f_N(T(x))$, $x \in \mathbb{R}^N$ defines a weakly differentiable function on $\mathbb{R}^N$ with $\left\|\hat{f}_N\right\|_{W^{1,2}(\mathbb{R}^N, \hat{q})} = 2^N \cdot N! \, \|f_N\|_{W^{1,2}(\Omega_N, \hat{q})}$. By [46, Theorem 2.5] the mollification with the standard mollifier yields an approximating sequence $\{u_l\}_l$ of smooth functions $u_l \in C_0^{\infty}(\mathbb{R}^N)$ of $\hat{f}_N$ in the weighted Sobolev spaces $H^{1,2}(\mathbb{R}^N, \hat{q})$. Hence, the restriction of the sequence $u_l \cdot \chi_N$ to Ω_N^{δ} approximates the restriction of $\hat{f}_N \cdot \chi_N$ to Ω_N^{δ} in $H^{1,2}(\Omega_N^{\delta}, \hat{q})$. Recall that $f_N = \hat{f}_N \cdot \chi_N$ on Ω_N^{δ}. Since χ_N is zero on $\overline{\Sigma}_N \setminus \Omega_N^{\delta}$ we obtain a sequence of $C^{\infty}(\overline{\Sigma}_N)$-functions $\{u_l \cdot \chi_N\}_l$ which converges to f_N in $H^{1,2}(\Sigma_N, q_N)$. This finishes the second step.

Step 3. In the third step we thus may assume w.l.o.g. that f_N is smooth up to the boundary of Σ_N. In particular, f_N is globally Lipschitz. Since f_N is integrable on Σ_N we may modify f_N close to the boundary to obtain a Lipschitz function $\tilde{f}_N$ which satisfies the Neumann condition and which is close to f_N in $\|.\|_1$. (Take, e.g. $\tilde{f}_N(x) = f(\pi(x))$, where $\pi(x)$ is the projection of x into the set $\Sigma_N^{\epsilon} = \{x \in \Sigma_N \,|\, \text{dist}(x, \partial\Sigma_N) \geq \epsilon\}$ for small $\epsilon > 0$.) We may now proceed as in step two to obtain an approximation of $\tilde{f}_N$ by smooth functions w.r.t. $\|.\|_1$, where we note that neither extension by reflection through the map T nor the standard mollification in [46] of the extended $\tilde{f}_N$ destroys the Neumann boundary condition. □

Corollary 3.25. *For quasi-every* $x \in \overline{\Sigma}_N$, *the the processes* $\tilde{X}^x_\cdot$ *and* $X^x_\cdot$ *are equal in law. In particular, X is a Feller process on* $\overline{\Sigma}_N$.

3.4 Semi-Martingale Properties

In this final section we prove the semi-martingale properties of X^N stated in Theorem 3.5. To that aim we establish the semi-martingale properties for the symmetric process X^N started in equilibrium, which imply the semi-martingale properties to hold for quasi-every starting point $x \in \overline{\Sigma}_N$ and by the Feller properties proven in the last section for every starting point $x \in \overline{\Sigma}_N$. In order to establish the semi-martingale properties of the stationary process, we shall use the following criterion established by Fukushima in [38]. For every open set $G \subset \overline{\Sigma}_N$ we set

$$\mathcal{C}_G := \{u \in D(\mathcal{E}^N) \cap C(\overline{\Sigma}_N) : \text{supp}(u) \subset G\}.$$

Theorem 3.26. *For* $u \in D(\mathcal{E}^N)$ *the additive functional* $u(X_t^N) - u(X_0^N)$ *is a semi-martingale if and only if one of the following (equivalent) conditions holds:*

i) For any relatively compact open set $G \subset \overline{\Sigma}_N$, there is a positive constant C_G such that

$$|\mathcal{E}^N(u, v)| \leq C_G \|v\|_{\infty}, \quad \forall v \in \mathcal{C}_G. \tag{3.18}$$

ii) There exists a signed Radon measure ν on $\overline{\Sigma}_N$ charging no set of zero capacity such that

$$\mathcal{E}^N(u,v) = -\int_{\overline{\Sigma}_N} v(x)\,\nu(dx), \qquad \forall v \in C(\overline{\Sigma}_N) \cap \mathcal{D}(\mathcal{E}^N). \tag{3.19}$$

Proof. See Theorem 6.3 in [38]. $\qquad\square$

Theorem 3.27. *Let X^N be a symmetric diffusion on $\overline{\Sigma}_N$ associated with the Dirichlet form $\mathcal{E}^N$, then X^N is an $\mathbb{R}^N$-valued semi-martingale if and only if $\beta/(N+1) \geq 1$.*

Proof. Since the semi-martingale property for $\mathbb{R}^N$-valued diffusions is defined componentwise, we shall apply Fukushima's criterion for $u(x) = x^i$, $i = 1, \ldots, N$.

Let us first consider the case where $\beta' := \beta/(N+1) > 1$. Then, for a relatively compact open set $G \subset \overline{\Sigma}_N$ and $v \in \mathcal{C}_G$,

$$\mathcal{E}^N(u,v) = \int_{\Sigma_N} \frac{\partial}{\partial x^i} v(x)\, q_N(dx)$$

$$= \frac{1}{Z_\beta} \int_0^1 dx^1 \int_{x^1}^1 dx^2 \cdots \int_{x^{i-1}}^1 dx^{i+1} \int_{x^{i+1}}^1 dx^{i+2} \cdots \int_{x^{N-1}}^1 dx^N \prod_{\substack{j=0\\ j\neq i-1,i}}^{N} (x^{j+1} - x^j)^{\beta'-1}$$

$$\times \int_{x^{i-1}}^{x^{i+1}} \frac{\partial}{\partial x^i} v(x)\, (x^i - x^{i-1})^{\beta'-1} (x^{i+1} - x^i)^{\beta'-1}\, dx^i.$$

Since $\beta' > 1$, we obtain by integration by parts

$$\int_{x^{i-1}}^{x^{i+1}} \frac{\partial}{\partial x^i} v(x)\, (x^i - x^{i-1})^{\beta'-1} (x^{i+1} - x^i)^{\beta'-1}\, dx^i$$

$$= -(\beta'-1) \int_{x^{i-1}}^{x^{i+1}} v(x) \left[(x^i - x^{i-1})^{\beta'-2}(x^{i+1} - x^i)^{\beta'-1} - (x^i - x^{i-1})^{\beta'-1}(x^{i+1} - x^i)^{\beta'-2} \right] dx^i$$

so that

$$\left| \int_{x^{i-1}}^{x^{i+1}} \frac{\partial}{\partial x^i} v(x)\, (x^i - x^{i-1})^{\beta'-1} (x^{i+1} - x^i)^{\beta'-1}\, dx^i \right|$$

$$\leq (\beta'-1)\|v\|_\infty \left(\int_{x^{i-1}}^{x^{i+1}} (x^i - x^{i-1})^{\beta'-2}\, dx^i + \int_{x^{i-1}}^{x^{i+1}} (x^{i+1} - x^i)^{\beta'-2}\, dx^i \right)$$

$$\leq 2(\beta'-1)\|v\|_\infty \int_0^1 r^{\beta'-2}\, dr \leq C\,\|v\|_\infty,$$

and we obtain that condition (3.18) holds. If $\beta' = 1$, i.e. the measure q_N coincides with the normalized Lebesgue measure on $\overline{\Sigma}_N$, condition (3.18) follows easily by a similar proceeding.

Let now $\beta' < 1$ and let us assume that $u(X_t^N) - u(X_0^N)$ is a semi-martingale. Then, there exists a signed Radon measure ν on $\overline{\Sigma}_N$ satisfying (3.19). Let $\nu = \nu_1 - \nu_2$ be the Jordan decomposition

3.4 Semi-Martingale Properties

of ν, i.e. ν_1 and ν_2 are positive Radon measures. By the above calculations we have for each relatively compact open set $G \subset \overline{\Sigma}_N$ and for all $v \in \mathcal{C}_G$

$$\mathcal{E}^N(u,v) = -\frac{1}{Z_\beta}(\beta'-1)\int_G v(x) \prod_{\substack{j=0 \\ j \neq i-1, i}}^{N} (x^{j+1}-x^j)^{\beta'-1}$$

$$\times \left[(x^i-x^{i-1})^{\beta'-2}(x^{i+1}-x^i)^{\beta'-1} - (x^i-x^{i-1})^{\beta'-1}(x^{i+1}-x^i)^{\beta'-2} \right] dx.$$

Hence, we obtain for the Jordan decomposition $\nu = \nu_1 - \nu_2$ that

$$\nu_1(G) = \frac{1}{Z_\beta}(1-\beta')\int_G (x^{i+1}-x^i)^{\beta'-2} \prod_{\substack{j=0 \\ j \neq i}}^{N}(x^{j+1}-x^j)^{\beta'-1} dx$$

$$\nu_2(G) = \frac{1}{Z_\beta}(1-\beta')\int_G (x^i-x^{i-1})^{\beta'-2} \prod_{\substack{j=0 \\ j \neq i-1}}^{N} (x^{j+1}-x^j)^{\beta'-1} dx.$$

Set $\partial\Sigma_N^j := \{x \in \partial\Sigma_N : x^j = x^{j+1}\}$, $j=0,\ldots,N$, and let for some $x_0 \in \partial\Sigma^i$ and $r > 0$, $A := x_0 + [-r,r]^N \cap \overline{\Sigma}_N$ be such that $\text{dist}(A, \partial\Sigma_N^j) > 0$ for all $j \neq i$. Furthermore, let $(A_n)_n$ be a sequence of compact subsets of A such that $A_n \uparrow A$ and $\text{dist}(A_n, \partial\Sigma_N^i) > 0$ for every n.

By the inner regularity of the Radon measures ν_1 and ν_2 we have $\nu_1(A) = \lim_n \nu_1(A_n)$ and $\nu_2(A) = \lim_n \nu_2(A_n)$. Since $\beta'-2 < -1$, we get by the choice of A that $\nu_1(A) = \infty$, while $\nu_2(A) < \infty$, which contradicts the local finiteness of ν and ν_1, respectively. $\square$

Chapter 4

Weak Convergence to the Wasserstein Diffusion

4.1 Introduction

The large scale behaviour of stochastic interacting particle systems is often described by linear or nonlinear deterministic evolution equations in the hydrodynamic limit, a fact which can be understood as a dynamic version of the law of large numbers, cf. e.g. [47]. Analogously, the fluctuations around such a deterministic limit usually lead to linear Ornstein-Uhlenbeck type stochastic partial differential equations (SPDE) on the diffusive time scale.

Here we add one more example to the collection of (in this case conservative) interacting particle systems with a nonlinear stochastic evolution in the hydrodynamic limit. Again we consider the process $(X^N) \in \overline{\Sigma}_N$ of N moving particles on the unit interval, which has been introduced and studied in the last chapter. Recall that $(X_t^N) = (x_t^1, \ldots, x_t^N)$ is a formal solution of the system

$$dx_t^i = (\frac{\beta}{N+1} - 1)\left(\frac{1}{x_t^i - x_t^{i-1}} - \frac{1}{x_t^{i+1} - x_t^i}\right)dt + \sqrt{2}dw_t^i + dl_t^{i-1} - dl_t^i, \quad i = 1, \cdots, N, \quad (4.1)$$

driven by independent real Brownian motions $\{w^i\}$ and local times l^i satisfying

$$dl_t^i \geq 0, \quad l_t^i = \int_0^t \mathbb{1}_{\{x_s^i = x_s^{i+1}\}} dl_s^i. \quad (4.2)$$

At first sight equation (4.1) resembles familiar Dyson-type models of interacting Brownian motions with electrostatic interaction. Except from the fact that (4.1) models a nearest-neighbour and not a mean-field interaction, the most important difference towards the Dyson model is however, that in the present case for $N \geq \beta - 1$ the drift is attractive and not repulsive. One technical consequence, as we have seen in the last chapter, is that the system (4.1) and (4.2) has to be understood properly because it may no longer be defined in the class of Euclidean semi-martingales. The second and more dramatic consequence is a clustering of hence strongly correlated particles such that fluctuations are seen on large hydrodynamic scales.

More precisely, we show that for properly chosen initial condition the empirical probability distribution of the particle system in the high density regime

$$\mu_t^N = \frac{1}{N}\sum_{i=1}^N \delta_{x_{N\cdot t}^i}$$

converges for $N \to \infty$ to the Wasserstein diffusion (μ_t) on the space of probability measures $\mathcal{P}([0,1])$, which was introduced in [66] as a conservative model for a diffusing fluid when its heat flow is perturbed by a kinetically uniform random forcing. In particular (μ_t) is a solution in the sense of an assocciated martingale problem to the SPDE

$$d\mu_t = \beta\Delta\mu_t dt + \Gamma(\mu_t)dt + \text{div}(\sqrt{2\mu_t}dB_t), \qquad (4.3)$$

where Δ is the Neumann Laplace operator and dB_t is space-time white noise over $[0,1]$ and, for $\mu \in \mathcal{P}([0,1])$, $\Gamma(\mu) \in \mathcal{D}'([0,1])$ is the Schwartz distribution acting on $f \in C^\infty([0,1])$ by

$$\langle \Gamma(\mu), f\rangle = \sum_{I\in \text{gaps}(\mu)}\left[\frac{f''(I_-)+f''(I_+)}{2} - \frac{f'(I_+)-f'(I_-)}{|I|}\right] - \frac{f''(0)+f''(1)}{2},$$

where $\text{gaps}(\mu)$ denotes the set of components $I =]I_-, I_+[$ of maximal length with $\mu(I) = 0$ and $|I|$ denotes the length of such a component.

The SPDE (4.3) has a familiar structure. For instance, the Dawson-Watanabe ('super-Brownian motion') process solves $d\mu_t = \Delta\mu_t dt + \sqrt{2\mu_t}dB_t$ whereas the empirical measure of a countable family of independent Brownian motions satifies the equation $d\mu_t = \Delta\mu_t dt + \text{div}(\sqrt{2\mu_t}dB_t)$, both again in the weak sense of the associated martingale problems, cf. e.g. [24].

The additional nonlinearity introduced through the operator Γ into (4.3) is crucial for the construction of (μ_t) by Dirichlet form methods because it guarantees the existence of a reversible measure $\mathbb{P}^\beta$ on $\mathcal{P}([0,1])$ which plays a central role for the convergence result, too. For $\beta > 0$, $\mathbb{P}^\beta$ can be defined as the law of the random probability measure $\eta \in \mathcal{P}([0,1])$ defined by

$$\langle f, \eta\rangle = \int_0^1 f(D_t^\beta)dt \quad \forall f \in C([0,1]),$$

where $t \to D_t^\beta = \frac{\gamma_{t\cdot\beta}}{\gamma_\beta}$ is the real valued Dirichlet process over $[0,1]$ with parameter β and γ denotes the standard Gamma subordinator.

It is argued in [66] that $\mathbb{P}^\beta$ admits the formal Gibbsean representation

$$\mathbb{P}^\beta(d\mu) = \frac{1}{Z}e^{-\beta\text{Ent}(\mu)}\mathbb{P}^0(d\mu)$$

with the Boltzmann entropy $\text{Ent}(\mu) = \int_{[0,1]}\log(d\mu/dx)d\mu$ as Hamiltonian and a particular uniform measure $\mathbb{P}^0$ on $\mathcal{P}([0,1])$, which illustrates the non-Gaussian character of $\mathbb{P}^\beta$. In particular, $\mathbb{P}^\beta$ is not log-concave. However an appropriate version of the Girsanov formula holds true for $\mathbb{P}^\beta$, see also [67], which implies the $L^2(\mathcal{P}([0,1]), \mathbb{P}^\beta)$-closability of the quadratic form

$$\mathcal{E}(F,F) = \int_{\mathcal{P}([0,1])}\|\nabla^w F\|_\mu^2 \, \mathbb{P}^\beta(d\mu), \quad F \in \mathcal{Z}$$

4.1 Introduction

on the class

$$\mathcal{Z} = \left\{ F : \mathcal{P}([0,1]) \to \mathbb{R} \ \middle| \ \begin{array}{l} F(\mu) = f(\langle \phi_1, \mu \rangle, \langle \phi_2, \mu \rangle, \ldots, \langle \phi_k, \mu \rangle) \\ f \in C_c^\infty(\mathbb{R}^k), \{\phi_i\}_{i=1}^k \subset C^\infty([0,1]), k \in \mathbb{N} \end{array} \right\}$$

where

$$\|\nabla^w F\|_\mu = \left\| (D_{|\mu} F)'(\cdot) \right\|_{L^2([0,1],\mu)}$$

and $(D_{|\mu} F)(x) = \partial_{t|t=0} F(\mu + t\delta_x)$. The corresponding closure, still denoted by $\mathcal{E}$, is a local regular Dirichlet form on the compact space $(\mathcal{P}([0,1]), \tau_w)$ of probabilities equipped with the weak topology. This allows to construct a unique Hunt diffusion process $((P_\eta)_{\eta \in \mathcal{P}([0,1])}, (\mu_t)_{t \geq 0})$ properly associated with $\mathcal{E}$, cf. [39]. Starting (μ_t) from equilibrium $\mathbb{P}^\beta$ yields what shall be called in the sequel exclusively the *Wasserstein diffusion* because its intrinsic metric coincides with the L^2-Wasserstein distance d_W, defined by

$$d_W(\mu, \nu) := \inf_\gamma \left(\iint_{[0,1]^2} |x-y|^2 \gamma(dx, dy) \right)^{1/2},$$

where the infimum is taken over all probability measures $\gamma \in \mathcal{P}([0,1]^2)$ having marginals μ and ν.

Our approach for the approximation result is again based on Dirichlet form methods. We use that the convergence of symmetric Markov semigroups is equivalent to an amplified notion of Gamma-convergence [52] of the associated sequence quadratic forms $\mathcal{E}^N$. In our situation it suffices to verify that equations (4.1) and (4.2) define a sequence of reversible finite dimensional particle systems whose equilibrium distributions converge nice enough to $\mathbb{P}^\beta$. In particular we show that also the logarithmic derivatives converge in an appropriate L^2-sense which implies the Mosco-convergence of the Dirichlet forms. (The pointwise convergence of the same sequence $\mathcal{E}^N$ to $\mathcal{E}$ has been used in a recent work by Döring and Stannat to establish the logarithmic Sobolev inequality for $\mathcal{E}$, cf. [27]). Since the approximating state spaces are finite dimensional we employ [50] for a generalized framework of Mosco convergence of forms defined on a scale of Hilbert spaces. In case of a fixed state space with varying reference measure the criterion of L^2-convergence of the associated logarithmic derivatives has been studied in e.g. [48]. However, in our case this result does not directly apply because in particular the metric and hence also the divergence operation itself is depending on the parameter N. However, only little effort is needed to see that things match up nicely, cf. Section 4.4.4.

A much more subtle point is the assumption of *Markov uniqueness* of the form $\mathcal{E}$ which we have to impose for the identification of the limit. By this we mean that $\mathcal{E}$ be a maximal element in the class of (not necessarily regular) Dirichlet forms on $L^2(\mathcal{P}([0,1]), \mathbb{P}^\beta)$ which is closely related to the Meyers-Serrin (*weak = strong*) property of the corresponding Sobolev space, cf. Corollary 4.20 and [31, 49]. Variants of this assumption appear in several quite similar infinite dimensional contexts as well [41, 48] and the verification depends crucially on the integration by parts formula which in the present case of $\mathbb{P}^\beta$ has a very peculiar structure. By general principles Markov uniqueness of $\mathcal{E}$ is weaker than the essential self-adjointness of the generator of $(\mu_t)_{t \geq 0}$ on $\mathcal{Z}_{\text{Neu}} = \{F \in \mathcal{Z} | \ F(\mu) = f(\langle \phi_1, \mu \rangle, \ldots, \langle \phi_k, \mu \rangle), \phi_i'(0) = \phi_i'(1) = 0, i = 1, \ldots, k\}$ and stronger than the well-posedness, i.e. uniqueness, in the class of Hunt processes on $\mathcal{P}([0,1])$ of the martingale problem defined by equation (4.3) on the set of test functions $\mathcal{Z}_{\text{Neu}}$, cf. [3, Theorem 3.4].

For the sake of a clearer presentation in the proofs we will work with a parametrization of a probability measure by the generalized right continuous inverse of its distribution function, which however is mathematically inessential. A side result of this parametrization is a diffusive scaling limit result for a (1+1)-dimensional gradient interface model with non-convex interaction potential (cf. [40]), see Section 4.5.

The material presented in this chapter is contained in [9].

4.2 Main Result

Let us briefly recall that the process $X_t^N = (x_t^1, \cdots, x_t^N)$ on the closure of $\Sigma_N := \{x \in \mathbb{R}^N : 0 < x^1 < x^2 < \cdots < x^N < 1\} \subset \mathbb{R}^N$, which can be seen as a generalized solution to the system (4.1) and (4.2), is a symmetric Markov process with equilibrium distribution

$$q_N(dx^1, \cdots, dx^N) = \frac{\Gamma(\beta)}{(\Gamma(\frac{\beta}{N+1}))^{N+1}} \prod_{i=1}^{N+1} (x^i - x^{i-1})^{\frac{\beta}{N+1}-1} dx^1 \ldots dx^N.$$

It can be rigorously constructed as the q_N-symmetric Hunt diffusion $((P_x)_{x \in \overline{\Sigma}_N}, (X_t^N)_{t \geq 0})$ associated to the local regular Dirichlet form $\mathcal{E}^N$ obtained as the $L^2(\Sigma_N, q_N)$-closure of

$$\mathcal{E}^N(f,g) = \int_{\Sigma_N} \nabla f(x) \cdot \nabla g(x) \, q_N(dx), \qquad f,g \in C^\infty(\overline{\Sigma}_N).$$

Now we can state the main result of this chapter.

Theorem 4.1. *For $\beta > 0$, assume that the Wasserstein Dirichlet form $\mathcal{E}$ is Markov unique. Let (X_t^N) denote the q_N-symmetric diffusion on $\overline{\Sigma}_N$ induced from the Dirichlet form $\mathcal{E}^N$, starting from equilibrium $X_0^N \sim q_N$, and let $\mu_t^N = \frac{1}{N} \sum_{i=1}^N \delta_{x_{N \cdot t}^i} \in \mathcal{P}([0,1])$, then the sequence of processes $(\mu_\cdot^N)$ converges weakly to $(\mu_\cdot)$ in $C_{\mathbb{R}_+}((\mathcal{P}([0,1]), \tau_w))$ for $N \to \infty$.*

4.3 Tightness

As usual we show compactness of the laws of $(\mu_\cdot^N)$ and, in a second step the uniqueness of the limit.

Proposition 4.2. *The sequence $(\mu_\cdot^N)$ is tight in $C_{\mathbb{R}_+}((\mathcal{P}([0,1]), \tau_w))$.*

Proof. According to Theorem 3.7.1 in [24] it is sufficient to show that the sequence $(\langle f, \mu_\cdot^N \rangle)_{N \in \mathbb{N}}$ is tight, where f is taken from a dense subset in $\mathcal{F} \subset C([0,1])$. Choose

$$\mathcal{F} := \{f \in C^3([0,1]) \, | \, f'(0) = f'(1) = 0\},$$

then $\langle f, \mu_t^N \rangle = F^N(X_{N \cdot t}^N)$ with

$$F^N(x) = \frac{1}{N} \sum_{i=1}^N f(x^i).$$

4.3 Tightness

The condition $f'(0) = f'(1) = 0$ implies $F^N \in C^2_{Neu}$ and

$$N \cdot L^N F^N(x) = \frac{\beta}{N+1} \sum_{i=1}^{N+1} \frac{f'(x^i) - f'(x^{i-1})}{x^i - x^{i-1}} - \sum_{i=1}^{N+1} \frac{f'(x^i) - f'(x^{i-1})}{x^i - x^{i-1}} + \sum_{i=1}^{N} f''(x^i),$$

which can be written as

$$N \cdot L^N F^N(x) = \frac{\beta}{N+1} \sum_{i=1}^{N+1} \frac{f'(x^i) - f'(x^{i-1})}{x^i - x^{i-1}}$$
$$+ \sum_{i=1}^{N} \left(f''(x^i) - \frac{f'(x^i) - f'(x^{i-1})}{x^i - x^{i-1}} \right) - \frac{f'(x^{N+1}) - f'(x^N)}{x^{N+1} - x^N}.$$

Finally, this can be estimated as follows:

$$|N \cdot L^N F^N(x)| \leq \|f''\|_\infty (\beta + 1) + \|f'''\|_\infty =: C(\beta, \|f\|_{C^3([0,1])}). \qquad (4.4)$$

This implies a uniform in N Lipschitz bound for the BV part in the Doob-Meyer decomposition of $F^N(X^N_{N\cdot})$. The process X^N has continuous sample paths with square field operator $\Gamma^N(F, F) = L^N(F^2) - 2F \cdot L^N F = |\nabla F|^2$. Hence the quadratic variation of the martingale part of $F^N(X^N_{N\cdot})$ satisfies

$$[F^N(X^N_{N\cdot})]_t - [F^N(X^N_{N\cdot})]_s = N \cdot \int_s^t |\nabla F^N|^2(X^N_s)\, ds = \frac{1}{N}\int_s^t \sum_{i=1}^{N}(f')^2(x^i_r)\, dr \leq (t-s)\, \|f'\|_\infty^2.$$
$$(4.5)$$

Since

$$F^N(X^N_0) = \frac{1}{N} \sum_{i=1}^{N} f(D^\beta_{i/(N+1)}) \longrightarrow \int_0^1 f(D^\beta_s)\, ds \qquad \mathbb{Q}^\beta\text{-a.s.,}$$

the law of $F^N(X^N_0)$ is convergent and by stationarity we conclude that also the law of $F^N(X^N_{N\cdot t})$ is convergent for every t. Using now Aldous' tightness criterion the assertion follows once we have shown that

$$\mathbb{E}\left[\left|F^N(X^N_{N\cdot(\tau_N+\delta_N)}) - F^N(X^N_{N\cdot\tau_N})\right|\right] \longrightarrow 0 \qquad \text{as } N \to \infty, \qquad (4.6)$$

for any given $\delta_N \downarrow 0$ and any given sequence of bounded stopping times (τ_N). Now, the Doob-Meyer decomposition reads as

$$F^N(X^N_{N\cdot(\tau_N+\delta_N)}) - F^N(X^N_{N\cdot\tau_N}) = M^N_{\tau_N+\delta_N} - M^N_{\tau_N} + \int_{\tau_N}^{\tau_N+\delta_N} N \cdot L^N F(X^N_{N\cdot s})\, ds,$$

where M^N is a martingale with quadratic variation $[M^N]_t = [F^N(X^N_{N\cdot})]_t$. Using (4.4) and (4.5) we get

$$\mathbb{E}\left[\left|F^N(X^N_{N\cdot(\tau_N+\delta_N)}) - F^N(X^N_{N\cdot\tau_N})\right|\right] \leq \mathbb{E}\left[\left|M^N_{\tau_N+\delta_N} - M^N_{\tau_N}\right|\right] + C_1\,\delta_N$$
$$\leq \mathbb{E}\left[\left|M^N_{\tau_N+\delta_N} - M^N_{\tau_N}\right|^2\right]^{1/2} + C_1\,\delta_N$$
$$= C_2\, \mathbb{E}\left[[M^N]_{\tau_N+\delta_N} - [M^N]_{\tau_N}\right]^{1/2} + C_1\,\delta_N$$
$$\leq C_3 \left(\delta_N^{1/2} + \delta_N\right),$$

for some positive constants C_i, which implies (4.6). □

The argument above shows the balance of first and second order parts of $N \cdot L^N$ as N tends to infinity. Alternatively one could use the symmetry of $(X_\cdot^N)$ and apply the Lyons-Zheng decomposition for the same result. We provide here some details.

For some fixed time $T > 0$ we use the Lyons-Zheng decomposition (see e.g. Theorem 5.7.1 in [39]) to obtain

$$F^N(X_{N \cdot t}^N) - F(X_0^N) = \frac{1}{2}\left(M_t - (\hat{M}_T - \hat{M}_{T-t})\right),$$

where M is a martingale w.r.t. the filtration $\mathcal{F}_t = \sigma(X_{N \cdot s}^N; 0 \leq s \leq t)$ and $\hat{M}$ is a martingale w.r.t. the filtration $\hat{\mathcal{F}}_t = \sigma(X_{N \cdot (T-s)}^N; 0 \leq s \leq t)$, $0 \leq t \leq T$. Moreover, the quadratic variation of M is given by

$$[M]_t = \int_0^t \Gamma^N(F^N, F^N)(X_{N \cdot s}^N)\, ds = N \cdot \int_0^t |\nabla F^N|^2(X_s^N)\, ds$$

and by symmetry we have $[\hat{M}]_t = [M]_t$, $0 \leq t \leq T$. Hence, we obtain using again (4.5)

$$\mathbb{E}\left[\left|F^N(X_{N \cdot t}^N) - F^N(X_{N \cdot s}^N)\right|\right] = \frac{1}{2}\mathbb{E}\left[|M_t - M_s|\right] + \frac{1}{2}\mathbb{E}\left[\left|\hat{M}_{T-t} - \hat{M}_{T-s}\right|\right]$$

$$\leq \frac{1}{2}\mathbb{E}\left[|M_t - M_s|^2\right]^{1/2} + \frac{1}{2}\mathbb{E}\left[\left|\hat{M}_{T-t} - \hat{M}_{T-s}\right|^2\right]^{1/2}$$

$$\leq \frac{1}{2}\mathbb{E}\left[|[M]_t - [M]_s|\right]^{1/2} + \frac{1}{2}\mathbb{E}\left[\left|[\hat{M}]_{T-t} - [\hat{M}]_{T-s}\right|\right]^{1/2}$$

$$\leq C|t - s|^{1/2},$$

for some positive constant C. Tightness follows now e.g. by Theorem 7.2 in Chapter 3 of [33].

4.4 Identification of the Limit

4.4.1 The $\mathcal{G}$-Parameterization

In order to identify the limit of the sequence $(\mu_\cdot^N)$ we parameterize the space $\mathcal{P}([0, 1])$ in terms of right continuous quantile functions, cf. e.g. [65, 66]. The set

$$\mathcal{G} = \{g : [0, 1) \to [0, 1] \mid g \text{ càdlàg nondecreasing}\},$$

equipped with the $L^2([0, 1], dx)$ distance d_{L^2} is a compact subspace of $L^2([0, 1], dx)$. It is homeomorphic to $(\mathcal{P}([0, 1]), \tau_w)$ by means of the map

$$\rho : \mathcal{G} \to \mathcal{P}([0, 1]), \qquad g \to g_*(dx),$$

which takes a function $g \in \mathcal{G}$ to the image measure of dx under g. The inverse map $\kappa = \rho^{-1} : \mathcal{P}([0, 1]) \to \mathcal{G}$ is realized by taking the right continuous quantile function, i.e.

$$g_\mu(t) := \inf\{s \in [0, 1] : \mu[0, s] > t\}.$$

For technical reasons we introduce the following modification of $(\mu_\cdot^N)$ which is better behaved in terms of the map κ.

4.4 IDENTIFICATION OF THE LIMIT

Lemma 4.3. *For $N \in \mathbb{N}$ define the Markov process*

$$\nu_t^N := \frac{N}{N+1}\mu_t^N + \frac{1}{N+1}\delta_0 \in \mathcal{P}([0,1]),$$

then $(\nu_\cdot^N)$ is convergent on $C_{\mathbb{R}_+}((\mathcal{P}([0,1]), \tau_w))$ if and only if $(\mu_\cdot^N)$ is. In this case both limits coincide.

Proof. Due to Theorem 3.7.1 in [24] it suffices to consider the sequences of real valued process $\langle f, \mu_\cdot^N \rangle$ and $\langle f, \nu_\cdot^N \rangle$ for $N \to \infty$, where $f \in C([0,1])$ is arbitrary. From

$$\text{i£} \cdot \sup_{t \geq 0} \left| \langle f, \mu_t^N \rangle - \langle f, \nu_t^N \rangle \text{i£} \cdot \right| = \sup_{t \geq 0} \frac{1}{N+1} \left| \langle f, \mu_t^N \rangle - \langle f, \delta_0 \rangle \right| \leq \frac{2}{N+1} \|f\|_\infty,$$

it follows that $d_C(\langle f, \mu_\cdot^N \rangle, \langle f, \nu_\cdot^N \rangle) \to 0$ almost surely, where d_C is any metric on $C([0,\infty), \mathbb{R})$ inducing the topology of locally uniform convergence. Hence for a bounded and uniformly d_C-continuous functional $F : C([0,\infty), \mathbb{R}) \to \mathbb{R}$

$$\text{i£} \cdot \mathbb{E}(F(\langle f, \mu_\cdot^N \rangle)) - \mathbb{E}(F(\langle f, \nu_\cdot^N \rangle)) \to 0 \quad \text{for } N \to \infty.$$

Since weak convergence on metric spaces is characterized by expectations of uniformly continuous, bounded functions (cf. e.g. [53, Theorem 6.1]) this proves the claim. □

Let $(g_\cdot^N) := (\kappa(\nu_\cdot^N))$ be the process $(\nu_\cdot^N)$ in the $\mathcal{G}$-parameterization. It can also be obtained by

$$g_t^N = \iota(X_{N \cdot t}^N)$$

with the imbedding $\iota = \iota^N$

$$\iota : \overline{\Sigma}_N \to \mathcal{G}, \qquad \iota(x) = \sum_{i=0}^{N} x^i \cdot \mathbb{1}_{[t_i, t_{i+1})},$$

with $t_i := i/(N+1)$, $i = 0, \ldots, N+1$. Similarly, let $(g_\cdot) = (\kappa(\mu_\cdot))$ be the $\mathcal{G}$-image of the Wasserstein diffusion under the map κ with invariant initial distribution $\mathbb{Q}^\beta$. In [66, Theorem 7.5] it is shown that $(g_\cdot)$ is generated by the Dirichlet form, again denoted by $\mathcal{E}$, which is obtained as the $L^2(\mathcal{G}, \mathbb{Q}^\beta)$-closure of

$$\mathcal{E}(u,v) = \int_\mathcal{G} \langle \nabla u_{|g}(\cdot), \nabla v_{|g}(\cdot) \rangle_{L^2([0,1])} \, \mathbb{Q}^\beta(dg), \qquad u,v \in \mathfrak{C}^1(\mathcal{G}).$$

on the class

$$\mathfrak{C}^1(\mathcal{G}) = \{u : \mathcal{G} \to \mathbb{R} \,|\, u(g) = U(\langle f_1, g \rangle_{L^2}, \ldots, \langle f_m, g \rangle_{L^2}), U \in C_c^1(\mathbb{R}^m), \{f_i\}_{i=1}^m \subset L^2([0,1]), m \in \mathbb{N}\},$$

where $\nabla u_{|g}$ is the $L^2([0,1], dx)$-gradient of u at g, defined via

$$\partial_s|_{s=0} u(g + s \cdot \xi) = \langle (\nabla_{|g} u)(\cdot), \xi(\cdot) \rangle_{L^2([0,1],dx)} \quad \forall \xi \in L^2([0,1], dx).$$

The convergence of $(\mu_\cdot^N)$ to $(\mu_\cdot)$ in $C_{\mathbb{R}_+}(\mathcal{P}([0,1]), \tau_w)$ is thus equivalent to the convergence of $(g_\cdot^N)$ to $(g_\cdot)$ in $C_{\mathbb{R}_+}(\mathcal{G}, d_{L^2})$. By Proposition 4.2 and Lemma 4.3 $(g_\cdot^N)_N$ is a tight sequence of processes on $\mathcal{G}$. The following statement identifies $(g_\cdot)$ as the unique weak limit.

Proposition 4.4. *Let $\mathcal{E}$ be Markov-unique. Then for any $f \in C(\mathcal{G}^l)$ and $0 \leq t_1 < \ldots < t_l$,*
$$\mathbb{E}(f(g_{t_1}^N, \cdots, g_{t_l}^N)) \xrightarrow{N \to \infty} \mathbb{E}(f(g_{t_1}, \cdots, g_{t_l})).$$

4.4.2 Finite Dimensional Approximation of Dirichlet Forms in Mosco Sense

Proposition 4.4 is proved by showing that the sequence of generating Dirichlet forms $N \cdot \mathcal{E}^N$ of (g_i^N) on $L^2(\Sigma_N, q_N)$ converges to $\mathcal{E}$ on $L^2(\mathcal{G}, \mathbb{Q})$ in the generalized Mosco sense of Kuwae and Shioya, allowing for varying base L^2-spaces. We recall the framework developed in [50].

Definition 4.5 (Convergence of Hilbert spaces). *A sequence of Hilbert spaces H^N converges to a Hilbert space H if there exists a family of linear maps $\{\Phi^N : H \to H^N\}_N$ such that*

$$\lim_N \left\| \Phi^N u \right\|_{H^N} = \|u\|_H, \quad \text{for all } u \in H.$$

A sequence $(u_N)_N$ with $u_N \in H_N$ converges strongly to a vector $u \in H$ if there exists a sequence $(\tilde{u}_N)_N \subset H$ tending to u in H such that

$$\lim_N \limsup_M \left\| \Phi^M \tilde{u}_N - u_M \right\|_{H^M} = 0,$$

and (u_N) converges weakly to u if

$$\lim_N \langle u_N, v_N \rangle_{H^N} = \langle u, v \rangle_H,$$

for any sequence $(v_N)_N$ with $v_N \in H^N$ tending strongly to $v \in H$. Moreover, a sequence $(B_N)_N$ of bounded operators on H^N converges strongly (resp. weakly) to an operator B on H if $B_N u_N \to Bu$ strongly (resp. weakly) for any sequence (u_N) tending to u strongly (resp. weakly).

Definition 4.6 (Mosco Convergence). *A sequence $(E^N)_N$ of quadratic forms E^N on H^N converges to a quadratic form E on H in the Mosco sense if the following two conditions hold:*

Mosco I: *If a sequence $(u_N)_N$ with $u_N \in H^N$ weakly converges to a $u \in H$, then*

$$E(u, u) \leq \liminf_N E^N(u_N, u_N).$$

Mosco II: *For any $u \in H$ there exists a sequence $(u_N)_N$ with $u_N \in H^N$ which converges strongly to u such that*

$$E(u, u) = \lim_N E^N(u_N, u_N).$$

Extending [52] it is shown in [50] that Mosco convergence of a sequence of Dirichlet forms is equivalent to the strong convergence of the associated resolvents and semigroups. We will apply this result when $H^N = L^2(\Sigma_N, q_N)$, $H = L^2(\mathcal{G}, \mathbb{Q}^\beta)$ and Φ^N is defined to be the conditional expectation operator

$$\Phi^N : H \to H^N; \quad (\Phi^N u)(x) := \mathbb{E}(u | g_{i/(N+1)} = x^i, i = 1, \ldots, N).$$

However, we shall prove that the sequence $N \cdot \mathcal{E}^N$ converges to $\mathcal{E}$ in the Mosco sense in a slightly modified fashion, namely the condition (Mosco II) will be replaced by

Mosco II': *There is a core $K \subset \mathcal{D}(E)$ such that for any $u \in K$ there exists a sequence $(u_N)_N$ with $u_N \in \mathcal{D}(E^N)$ which converges strongly to u such that $E(u,u) = \lim_N E^N(u_N, u_N)$.*

4.4 IDENTIFICATION OF THE LIMIT

Theorem 4.7. *Under the assumption that $H^N \to H$ the conditions (Mosco I) and (Mosco II') are equivalent to the strong convergence of the associated resolvents.*

Proof. We proceed as in the proof of Theorem 2.4.1 in [52]. By Theorem 2.4 of [50] strong convergence of resolvents implies Mosco-convergence in the original stronger sense. Hence we need to show only that our weakened notion of Mosco-convergence also implies strong convergence of resolvents.

Let $\{R_\lambda^N, \lambda > 0\}$ and $\{R_\lambda, \lambda > 0\}$ be the resolvent operators associated with E^N and E, respectively. Then, for each $\lambda > 0$ we have to prove that for every $z \in H$ and every sequence (z_N) tending strongly to z the sequence (u_N) defined by $u_N := R_\lambda^N z_N \in H^N$ converges strongly to $u := R_\lambda z$ as $N \to \infty$. The vector u is characterized as the unique minimizer of $E(v,v) + \lambda \langle v, v \rangle_H - 2\langle z, v \rangle_H$ over H and a similar characterization holds for each u_N. Since for each N the norm of R_λ^N as an operator on H^N is bounded by λ^{-1}, by Lemma 2.2 in [50] there exists a subsequence of (u_N), still denoted by (u_N), that converges weakly to some $\tilde{u} \in H$. By (Mosco II') we find for every $v \in K$ a sequence (v_N) tending strongly to v such that $\lim_N E^N(v_N, v_N) = E(v, v)$. Since for every N

$$E^N(u_N, u_N) + \lambda \langle u_N, u_N \rangle_{H^N} - 2\langle z_N, u_N \rangle_{H^N} \leq E^N(v_N, v_N) + \lambda \langle v_N, v_N \rangle_{H^N} - 2\langle z_N, v_N \rangle_{H^N},$$

using the condition (Mosco I) we obtain in the limit $N \to \infty$:

$$E(\tilde{u}, \tilde{u}) + \lambda \langle \tilde{u}, \tilde{u} \rangle_H - 2\langle z, \tilde{u} \rangle_H \leq E(v, v) + \lambda \langle v, v \rangle_H - 2\langle z, v \rangle_H,$$

which by the definition of the resolvent together with the density of $K \subset D(E)$ implies that $\tilde{u} = R_\lambda z = u$. This establishes the weak convergence of resolvents. It remains to show strong convergence. Let $u_N = R_\lambda^N z_N$ converge weakly to $u = R_\lambda z$ and choose $v \in K$ with the respective strong approximations $v_N \in H^N$ such that $E^N(v_N, v_N) \to E(v, v)$, then the resolvent inequality for R_λ^N yields

$$E^N(u_N, u_N) + \lambda \|u_N - z_N/\lambda\|_{H^N}^2 \leq E^N(v_N, v_N) + \lambda \|v_N - z_N/\lambda\|_{H^N}^2.$$

Taking the limit for $N \to \infty$, one obtains

$$\limsup_N \lambda \|u_N - z_N/\lambda\|_{H^N}^2 \leq E(v, v) - E(u, u) + \lambda \|v - z/\lambda\|_H^2.$$

Since K is a dense subset we may now let $v \to u \in D(E)$, which yields

$$\limsup_N \|u_N - z_N/\lambda\|_{H^N}^2 \leq \|u - z/\lambda\|_H^2.$$

Due to the weak lower semicontinuity of the norm this yields $\lim_N \|u_N - z_N/\lambda\|_{H^N} = \|u - z/\lambda\|_H$. Since strong convergence in H is equivalent to weak convergence together with the convergence of the associated norms the claim follows (cf. Lemma 2.3 in [50]). □

Proposition 4.4 will now essentially be implied by the following statement, which will be proven in the next two subsections.

Theorem 4.8. *Assume that $\mathcal{E}$ is Markov-unique on $L^2(\mathcal{G},\mathbb{Q})$. Then $(N \cdot \mathcal{E}^N, H^N)$ converges to $(\mathcal{E}, H)$ along Φ^N in Mosco sense.*

Proposition 4.9. H^N *converges to H along Φ^N, for $N \to \infty$.*

Proof. We have to show that $\left\|\Phi^N u\right\|_{H^N} \to \|u\|_H$ for each $u \in H$. Let $\mathcal{F}^N$ be the σ-algebra on $\mathcal{G}$ generated by the projection maps $\{g \to g(i/(N+1)) \mid i=1,\ldots,N\}$. By abuse of notation we identify $\Phi^N u \in H^N$ with $\mathbb{E}(u|\mathcal{F}^N)$ of u, considered as an element of $L^2(\mathbb{Q}^\beta, \mathcal{F}^N) \subset H$. Since the measure q_N coincides with the respective finite dimensional distributions of $\mathbb{Q}^\beta$ on $\overline{\Sigma}_N$ we have $\left\|\Phi^N u\right\|_{H^N} = \left\|\Phi^N u\right\|_H$. Hence the claim will follow once we show that $\Phi^N u \to u$ in H. For the latter we use the following abstract result, whose proof can be found, e.g. in [4, Lemma 1.3].

Lemma 4.10. *Let $(\Omega, \mathcal{D}, \mu)$ be a measure space and $(\mathcal{F}_n)_{n \in \mathbb{N}}$ a sequence of σ-subalgebras of $\mathcal{D}$. Then $E(f|\mathcal{F}_n) \to f$ for all $f \in L^p$, $p \in [1,\infty)$ if and only if for all $A \in \mathcal{D}$ there is a sequence $A_n \in \mathcal{F}_n$ such that $\mu(A_n \Delta A) \to 0$ for $n \to \infty$.*

In order to apply this lemma to the given case $(\mathcal{G}, \mathcal{B}(\mathcal{G}), \mathbb{Q}^\beta)$, where $\mathcal{B}(\mathcal{G})$ denotes the Borel σ-algebra on $\mathcal{G}$, let $\mathcal{F}_{\mathbb{Q}^\beta} \subset \mathcal{B}(\mathcal{G})$ denote the collection of all Borel sets $F \subset \mathcal{G}$ which can be approximated by elements $F_N \in \mathcal{F}^N$ with respect to $\mathbb{Q}^\beta$ in the sense above. Note that $\mathcal{F}_{\mathbb{Q}^\beta}$ is again a σ-algebra, cf. the appendix in [4]. Let $\mathcal{M}$ denote the system of finitely based open cylinder sets in $\mathcal{G}$ of the form $M = \{g \in \mathcal{G} | g_{t_i} \in O_i, i=1,\ldots,L\}$ where $t_i \in [0,1]$ and $O_i \subset [0,1]$ open. From the almost sure right continuity of g and the fact that $g.$ is continuous at $t_1,\ldots,t_L$ for $\mathbb{Q}^\beta$-almost all g it follows that $M_N := \{g \in \mathcal{G} | g_{\lceil t_i \cdot (N+1)\rceil/(N+1)} \in O_i, i=1,\ldots,L\} \in \mathcal{F}^N$ is an approximation of M in the sense above. Since $\mathcal{M}$ generates $\mathcal{B}(\mathcal{G})$ we obtain $\mathcal{B}(\mathcal{G}) \subset \mathcal{F}_{\mathbb{Q}^\beta}$ such that the assertion holds, due to Lemma 4.10. □

Remark 4.11. It is much simpler to prove Proposition 4.9 for a dyadic subsequence $N' = 2^m - 1$, $m \in \mathbb{N}$ when the sequence $\left\|\Phi^{N'} u\right\|_{H^{N'}}$ is nondecreasing and bounded, because $\Phi^{N'}$ is a projection operator in H with increasing range $\text{im}(\Phi^{N'})$ as N' grows. Hence, $\left\|\Phi^{N'} u\right\|_{H^{N'}}$ is Cauchy and thus

$$\left\|\Phi^{N'} u - \Phi^{M'} u\right\|_H^2 = \left\|\Phi^{N'} u\right\|_H^2 - \left\|\Phi^{M'} u\right\|_H^2 \to 0 \quad \text{for } M', N' \to \infty,$$

i.e. the sequence $\Phi^{N'} u$ converges to some $v \in H$. Since obviously $\Phi^N u \to u$ weakly in H it follows that $u = v$ such that the claim is obtained from $\left| \left\|\Phi^{N'} u\right\|_H - \|u\|_H \right| \leq \left\|\Phi^{N'} u - u\right\|_H$.

4.4.3 Condition Mosco II'

To simplify notation for $f \in L^2([0,1], dx)$ denote the functional $g \to \langle f, g \rangle_{L^2([0,1])}$ on $\mathcal{G}$ by l_f. We introduce the set K of polynomials defined by

$$K = \left\{ u \in C(\mathcal{G}) \mid u(g) = \prod_{i=1}^n l_{f_i}^{k_i}(g),\, k_i \in \mathbb{N},\, f_i \in C([0,1]) \right\}.$$

Lemma 4.12. *The linear span of K is a core of $\mathcal{E}$.*

4.4 IDENTIFICATION OF THE LIMIT

Proof. Recall that by [66, Theorem 7.5] the set

$$\mathfrak{C}^1(\mathcal{G}) = \{u : \mathcal{G} \to \mathbb{R} \mid u(g) = U(\langle f_1, g \rangle_{L^2}, \ldots, \langle f_m, g \rangle_{L^2}), U \in C_c^1(\mathbb{R}^m), \{f_i\}_{i=1}^m \subset L^2([0,1]), m \in \mathbb{N}\},$$

is a core for the Dirichlet form $\mathcal{E}$ in the $\mathcal{G}$-parametrization. The boundedness of $\mathcal{G} \subset L^2([0,1], dx)$ implies that U is evaluated on a compact subset of $\mathbb{R}^m$ only, where U can be approximated by polynomials in the C^1-norm. From this, the chain rule for the L^2-gradient operator ∇ and Lebesgue's dominated convergence theorem in $L^2(\mathcal{G}, \mathbb{Q}^\beta)$ it follows that the linear span of polynomials of the form $u(g) = \prod_{i=1}^n l_{f_i}^{k_i}(g)$ with $k_i \in \mathbb{N}$, $f_i \in C([0,1])$, $k_i \in \mathbb{N}$, is also a core of $\mathcal{E}$. □

Lemma 4.13. *For a polynomial $u \in K$ with $u(g) = \prod_{i=1}^n l_{f_i}^{k_i}(g)$ let $u_N := \prod_{i=1}^n (\Phi^N(l_{f_i}))^{k_i} \in H^N$, then $u_N \to u$ strongly.*

Proof. Let $\tilde{u}_N := \prod_{i=1}^n (\Phi^N(l_{f_i}))^{k_i} \in H$ be the respective product of conditional expectations, where as above Φ^N also denotes the projection operator on $H = L^2(\mathcal{G}, \mathbb{Q}^\beta)$. Note that by Jensen's inequality for any measurable functional $u : \mathcal{G} \to \mathbb{R}$, $|\Phi^N(u)|(g) \leq \Phi^N(|u|)(g)$ for $\mathbb{Q}^\beta$-almost all $g \in G$, such that in particular $\left\|\Phi^N(l_{f_i})\right\|_{L^\infty(\mathcal{G}, \mathbb{Q}^\beta)} \leq \|l_{f_i}\|_{L^\infty(\mathcal{G}, \mathbb{Q}^\beta)} \leq \|f_i\|_{C([0,1])}$. Hence each of the factors $\Phi^N(l_{f_i}) \in H$ is uniformly bounded and converges strongly to l_{f_i} in $L^2(\mathcal{G}, \mathbb{Q}^\beta)$, such that the convergence also holds true in any $L^p(\mathcal{G}, \mathbb{Q}^\beta)$ with $p > 0$. This implies $\tilde{u}_N \to u$ in H. Furthermore,

$$\lim_N \lim_M \left\|\Phi^M \tilde{u}_N - u_M\right\|_{H^M} = \lim_N \lim_M \left\|\Phi^M\left(\prod_{i=1}^n (\Phi^N(l_{f_i}))^{k_i}\right) - \prod_{i=1}^n (\Phi^M(l_{f_i}))^{k_i}\right\|_H$$

$$= \lim_N \left\|\prod_{i=1}^n (\Phi^N(l_{f_i}))^{k_i} - \prod_{i=1}^n l_{f_i}^{k_i}\right\|_H = 0. \quad (4.7)$$

□

For the proof of Mosco II' we will also need that the conditional expectation of the random variable g w.r.t. to $\mathbb{Q}^\beta$ given finitely many intermediate points $\{g(t_i) = x^i\}$ yields the linear interpolation.

Lemma 4.14. *For $X \in \overline{\Sigma}_N$ define $g_X \in \mathcal{G}$ by*

$$g_X(t) = x^i + ((N+1) \cdot t - i)(x^{i+1} - x^i) \quad \text{if } t \in [\frac{i}{N+1}, \frac{i+1}{N+1}), \quad i = 0, \ldots, N,$$

then

$$\mathbb{E}(g|\mathcal{F}^N)(X) = g_X.$$

Proof. This quite classical claim follows essentially from the bridge representation of the Dirichlet process, i.e. $\mathbb{Q}^\beta$ is the law of $(\gamma(\beta \cdot t)_{t \in [0,1]} \in G)$ on $\mathcal{G}$ conditioned on $\gamma(\beta) = 1$ where γ is the standard Gamma subordinator, cf. e.g. [66]. Together with the elementary property that $\mathbb{E}_{\mathbb{Q}^\beta}(g(t)) = t$ for $t \in [0,1]$ the claim follows from the homogeneity of γ together with simple scaling and iterated use of the Markov property. See Appendix A for more details. □

Proposition 4.15. *For all $u \in K$ there is a sequence $u_N \in \mathcal{D}(\mathcal{E}^N)$ converging strongly to u in H and $N \cdot \mathcal{E}^N(u_N, u_N) \to \mathcal{E}(u, u)$. In particular, condition Mosco II' is satisfied.*

Proof. For $u \in K$ let $u_N := \prod_{i=1}^{n}(\Phi^N(l_{f_i}))^{k_i} \in H^N$ as above then the strong convergence of u^N to u is assured by Lemma 4.13. From Lemma 4.14 we obtain that $\Phi^N(l_f)(X) = \langle f, g_X \rangle$. In particular

$$(\nabla \Phi^N(l_f)(X))^i = \frac{1}{N+1} \cdot (\eta^N * f)(t_i), \qquad (4.8)$$

where $t_i := i/(N+1)$, $i = 1, \ldots, N+1$ and η^N denotes the convolution kernel $t \to \eta^N(t) = (N+1) \cdot (1 - \min(1, |(N+1) \cdot t|))$, which can be easily seen as follows: Setting $t_j := j/(N+1)$, $j \in \{0, \ldots, N+1\}$, we have on one hand

$$\Phi^N(l_f)(X) = \int_0^1 f(t) g_X(t) \, dt$$

$$= \sum_{j=0}^{N} \int_{t_j}^{t_{j+1}} f(t) \left(x^j + ((N+1)t - j)(x^{j+1} - x^j) \right) dt$$

$$= \sum_{j=0}^{N} \int_{t_j}^{t_{j+1}} f(t) \left(1 - ((N+1)t - j) \right) x^j \, dt + \sum_{j=0}^{N} \int_{t_j}^{t_{j+1}} f(t) \left((N+1)t - j \right) x^{j+1} \, dt,$$

so that

$$(\nabla \Phi^N(l_f)(X))^i = \int_{t_i}^{t_{i+1}} f(t) \left(1 - ((N+1)t - i) \right) dt + \int_{t_{i-1}}^{t_i} f(t) \left((N+1)t - (i-1) \right) dt$$

for all $i \in \{1, \ldots, N\}$. On the other hand

$$\frac{1}{N+1} \cdot (\eta^N * f)(t_i) = \frac{1}{N+1} \int_{\mathbb{R}} f(t) \, \eta^N(t_i - t) \, dt = \int_{\mathbb{R}} f(t) \left(1 - \min(1, |i - (N+1)t|) \right) dt.$$

Since

$$1 - \min(1, |i - (N+1)t|) = \begin{cases} 0 & \text{if } t \leq t_{i-1} \text{ or } t \geq t_{i+1}, \\ 1 - (i - (N+1)t) & \text{if } t_{i-1} \leq t \leq t_i, \\ 1 + (i - (N+1)t) & \text{if } t_i \leq t \leq t_{i+1}, \end{cases}$$

we obtain (4.8).

For the convergence of $N \cdot \mathcal{E}_N(u^N, u^N)$ to $\mathcal{E}(u, u)$ let $f \in C([0,1])$ be arbitrary, in particular f is uniformly continuous. Since $\int_{\mathbb{R}} \eta^N(t) \, dt = 1$ and $\eta^N = 0$ on $[-\frac{1}{N+1}, \frac{1}{N+1}]^c$ we have

$$\min_{s \in \left[t - \frac{1}{N+1}, t + \frac{1}{N+1}\right]} f(s) \leq \int_{\mathbb{R}} f(t - s) \, \eta^N(s) \, ds \leq \max_{s \in \left[t - \frac{1}{N+1}, t + \frac{1}{N+1}\right]} f(s),$$

4.4 IDENTIFICATION OF THE LIMIT

and from the uniform continuity of f we can conclude that $f * \eta^N \to f$ in $C([0,1])$ as $N \to \infty$. Hence, using (4.8) we also get

$$N \cdot |\nabla \Phi^N(l_f)(X)|^2 = \frac{N}{(N+1)^2} \sum_{i=1}^{N} (\eta^N * f)^2(t_i)$$

$$= \frac{N}{(N+1)^2} \sum_{i=1}^{N} \left[(\eta^N * f)^2(t_i) - f^2(t_i) \right] + \frac{N}{(N+1)^2} \sum_{i=1}^{N} f^2(t_i)$$

$$\longrightarrow \langle f, f \rangle_{L^2([0,1])}.$$

as $N \to \infty$. Since the gradient does not depend on the value of the vector X and $\nabla_{L^2} l_f = f$, this implies the claim in the case when $u = l_f$. Similarly, for arbitrary $f_1, f_2 \in C([0,1])$

$$N \cdot \left\langle \nabla \Phi^N(l_{f_1}), \nabla \Phi^N(l_{f_2}) \right\rangle_{\mathbb{R}^N} \longrightarrow \langle f_1, f_2 \rangle_{L^2([0,1])}. \tag{4.9}$$

Consider now $u \in K$ with $u(g) = \prod_{i=1}^{n} l_{f_i}^{k_i}(g)$. The chain rule for the L^2-gradient operator $\nabla = \nabla^{L^2}$ yields

$$\nabla u(g) = \sum_{j=1}^{n} \left(\prod_{i \neq j} l_{f_i}^{k_i}(g) \right) k_j \, l_{f_j}^{k_j - 1}(g) \, f_j.$$

Thus,

$$\langle \nabla u, \nabla u \rangle_{L^2(0,1)} = \sum_{j,s=1}^{n} \left(\prod_{i \neq j} l_{f_i}^{k_i} \right) \left(\prod_{r \neq s} l_{f_r}^{k_r} \right) k_j \, k_s \, l_{f_j}^{k_j - 1} l_{f_s}^{k_s - 1} \langle f_j, f_s \rangle_{L^2(0,1)}.$$

and analogously for $u_N = \prod_{i=1}^{n} (\Phi^N(l_{f_i}))^{k_i}$ with $\nabla = \nabla^{\mathbb{R}^N}$

$$\nabla u_N = \sum_{j=1}^{n} \left(\prod_{i \neq j} (\Phi^N(l_{f_i}))^{k_i} \right) k_j \, (\Phi^N(l_{f_j}))^{k_j - 1} \nabla \Phi^N(l_{f_j})$$

and

$$\langle \nabla u_N, \nabla u_N \rangle_{\mathbb{R}^N} = \sum_{j,s=1}^{n} \left(\prod_{i \neq j} (\Phi^N(l_{f_i}))^{k_i} \right) \left(\prod_{r \neq s} (\Phi^N(l_{f_r}))^{k_r} \right)$$

$$\times k_j \, k_s \, (\Phi^N(l_{f_j}))^{k_j - 1} (\Phi^N(l_{f_s}))^{k_s - 1} \left\langle \nabla \Phi^N(l_{f_j}), \nabla \Phi^N(l_{f_s}) \right\rangle_{\mathbb{R}^N}.$$

Since

$$N \cdot \mathcal{E}^N(u_N, u_N) = N \cdot \int_{\Sigma_N} \langle \nabla u_N, \nabla u_N \rangle_{\mathbb{R}^N} dq_N$$

$$= N \cdot \int_G \langle \nabla u_N, \nabla u_N \rangle (g(t_1), \ldots, g(t_N)) \, \mathbb{Q}^\beta(dg)$$

and for $\mathbb{Q}^\beta$-a.e. g

$$\Phi^N(l_f)(g(t_1), \ldots, g(t_N)) \longrightarrow l_f(g) \quad \text{as } N \to \infty,$$

if $f \in C([0,1])$ the first assertion of the proposition holds by (4.9) and dominated convergence. The second assertion follows now from linearity and polarisation together with Lemma 4.12. □

For later use we make an observation which follows easily from the proof of the last proposition.

Lemma 4.16. *For u and u_N as in the proof of Proposition 4.15 and for $\mathbb{Q}^\beta$-a.e. g we have*

$$\left\|(N+1)\iota^N(\nabla u_N(g(t_1),\ldots,g(t_N))) - \nabla u_{|g}\right\|_{L^2(0,1)} \longrightarrow 0 \quad \text{as } N \to \infty,$$

with $\iota^N : \mathbb{R}^N \to D([0,1), \mathbb{R})$ defined as above and $t_l := l/(N+1)$, $l = 0, \ldots, N+1$.

Proof. By the definitions we have for every $x \in \overline{\Sigma}_N$

$$(N+1)\iota^N(\nabla u_N(x)) = \sum_{j=1}^n \left(\prod_{i \neq j}(\Phi^N(l_{f_i})(x))^{k_i}\right) k_j \left(\Phi^N(l_{f_j})(x)\right)^{k_j-1}$$

$$\times (N+1)\sum_{l=1}^N \nabla(\Phi^N(l_{f_j})(x))^l \mathbb{1}_{[t_l, t_{l+1}]}$$

$$= \sum_{j=1}^n \left(\prod_{i \neq j}(\Phi^N(l_{f_i})(x))^{k_i}\right) k_j \left(\Phi^N(l_{f_j})(x)\right)^{k_j-1} \sum_{l=1}^N (\eta^N * f_j)(t_l) \mathbb{1}_{[t_l, t_{l+1}]},$$

where we have used again (4.8). Furthermore, for every j,

$$\int_0^1 \left(\sum_{l=1}^N (\eta^N * f_j)(t_l) \mathbb{1}_{[t_l, t_{l+1}]}(t) - f_j(t)\right)^2 dt$$

$$= \sum_{l=1}^N \int_{t_l}^{t_{l+1}} \left((\eta^N * f_j)(t_l) - f_j(t)\right)^2 dt$$

$$\leq 2 \sum_{l=1}^N \int_{t_l}^{t_{l+1}} \left((\eta^N * f_j)(t_l) - f_j(t_l)\right)^2 dt + 2 \sum_{l=1}^N \int_{t_l}^{t_{l+1}} (f_j(t_l) - f_j(t))^2 dt,$$

where the first term tends to zero as $N \to \infty$ since $\eta^N * f_j \to f_j$ in the sup-norm and the second term tends to zero by the uniform continuity of f_j. From this we can directly deduce the claim because $\Phi^N(l_f)(g(t_1), \ldots, g(t_N)) \to l_f(g)$ as $N \to \infty$ for $\mathbb{Q}^\beta$-a.e. g. □

4.4.4 Condition Mosco I

For the verification of Mosco I we exploit that the respective integration by parts formulas of $\mathcal{E}^N$ and $\mathcal{E}$ converge. In case of a fixed state space a similar approach is discussed abstractly in [48]. However, here also the state spaces of the processes change which requires some extra care for the varying metric structures in the Dirichlet forms.

Let $T^N := \{f : \overline{\Sigma}_N \to \mathbb{R}^N\}$ be equipped with the norm

$$\|f\|_{T^N}^2 := \frac{1}{N} \int_{\Sigma_N} \|f(x)\|_{\mathbb{R}^N}^2 \, q_N(dx),$$

then the corresponding integration by parts formula for q_N on Σ_N, established in Proposition 3.7, reads

$$\langle \nabla u, \xi \rangle_{T^N} = -\frac{1}{N} \langle u, \operatorname{div}_{q_N} \xi \rangle_{H^N}. \tag{4.10}$$

4.4 IDENTIFICATION OF THE LIMIT

To state the corresponding formula for $\mathcal{E}$ we introduce the Hilbert space of vector fields on $\mathcal{G}$ by
$$T = L^2(\mathcal{G} \times [0,1], \mathbb{Q}^\beta \otimes dx),$$
and the subset $\Theta \subset T$
$$\Theta = \mathrm{span}\{\zeta \in T \mid \zeta(g,t) = w(g) \cdot \varphi(g(t)), w \in K, \varphi \in \mathcal{C}^\infty([0,1]) : \varphi(0) = \varphi(1) = 0\}.$$

Lemma 4.17. Θ *is dense in* T.

Proof. Let us first remove the condition $\phi(0) = \phi(1)$, i.e let us show that the T-closure of Θ coincides with that of $\bar\Theta = \mathrm{span}\{\zeta \in T \mid \zeta(g,t) = w(g) \cdot \varphi(g(t)), w \in K, \varphi \in \mathcal{C}^\infty([0,1])\}$. Since $\sup_{g \in \mathcal{G}} w(g) < \infty$ for any $w \in K$, it suffices to show that any $\zeta \in \bar\Theta$ of the form $\zeta(g,t) = \varphi(g(t))$ can be approximated in T by functions $\zeta_k(g,t) = \varphi_k(g(t))$ with $\varphi_k \in C^\infty([0,1])$ and $\varphi_k(0) = \varphi_k(1) = 0$. Choose a sequence of functions $\varphi_k \in C_0^\infty([0,1])$ such that such that $\sup_k \|\varphi_k\|_{C([0,1])} < \infty$ and $\varphi_k(s) \to \varphi(s)$ for all $s \in]0,1[$. Now for $\mathbb{Q}^\beta$-almost all g it holds that $\{s \in [0,1] \mid g(s) = 0\} = \{0\}$ and $\{s \in [0,1] \mid g(s) = 1\} = \{1\}$, such that $\phi_k(g(s)) \to \phi(g(s))$ for $\mathbb{Q}^\beta \otimes dx$ -almost all (g,s). The uniform boundedness of the sequence of functions $\zeta_k : \mathcal{G} \times [0,1] \to \mathbb{R}$ together with dominated convergence w.r.t. the measure $\mathbb{Q}^\beta \otimes dx$ the convergence is established. In order to complete the proof of the lemma note that $\mathbb{Q}^\beta$-amost every $g \in G$ is a strictly increasing funcion on $[0,1]$. This implies that the set $\bar\Theta$ is separating the points of a full measure subset of $\mathcal{G} \times [0,1]$. Hence the assertion follows from the following abstract lemma. □

Lemma 4.18. *Let μ be a probability measure on a Polish space (X,d) and let $\mathcal{A}$ be a subalgebra of $C(X)$ containing the constants. Assume that $\mathcal{A}$ is μ-almost everywhere separating on X, i.e. there exists a measurable subset $\tilde X$ with $\mu(\tilde X) = 1$ and for all $x,y \in \tilde X$ there is an $a \in \mathcal{A}$ such that $a(x) \neq a(y)$. Then $\mathcal{A}$ is dense in any $L^p(X,\mu)$ for $p \in [1, \infty)$.*

Proof. We may assume w.l.o.g. that $\mathcal{A}$ is stable w.r.t. the operation of taking the pointwise inf and sup. Let $u \in L^p(X)$, then we may also assume w.l.o.g. that u is continuous and bounded on X. By the regularity of μ we can approximate $\tilde X$ from inside by compact subsets K_m such that $\mu(K_m) \geq 1 - \frac{1}{m}$. On each K_m the theorem of Stone-Weierstrass tells that there is some $a_m \in \mathcal{A}$ such that $\left\|u_{|K_m} - a_{m|K_m}\right\|_{C(K_m)} \leq \frac{1}{m}$, and by truncation $\|a_m\|_{C(X)} \leq \|u\|_{C(X)}$. In particular, for $\epsilon > 0$, $\mu(|a_m - u| > \epsilon) \leq \mu(X \setminus K_m) \leq \frac{1}{m}$, if $m \geq 1/\epsilon$, i.e. a_m converges to u on X in μ-probability. Hence some subsequence $a_{m'}$ converges to u pointwise μ-a.s. on X, and hence the claim follows from the uniform boundedness of the a_m by dominated convergence. □

The L^2-derivative operator ∇ defines a map
$$\nabla : \mathfrak{C}^1(\mathcal{G}) \to T$$
which by [66, Proposition 7.3], cf. [67], satisfies the following integration by parts formula.
$$\langle \nabla u, \zeta \rangle_T = -\langle u, \mathrm{div}_{\mathbb{Q}^\beta} \zeta \rangle_H, \quad u \in \mathfrak{C}^1(\mathcal{G}), \zeta \in \Theta, \tag{4.11}$$

where, for $\zeta(g,t) = w(g) \cdot \varphi(g(t))$,

$$\text{div}_{\mathbb{Q}^\beta} \zeta(g) = w(g) \cdot V_\varphi^\beta(g) + \langle \nabla w(g)(.), \varphi(g(.)) \rangle_{L^2(dx)}$$

with

$$V_\varphi^\beta(g) := V_\varphi^0(g) + \beta \int_0^1 \varphi'(g(x))dx - \frac{\varphi'(0) + \varphi'(1)}{2}$$

and

$$V_\varphi^0(g) := \sum_{a \in J_g} \left[\frac{\varphi'(g(a+)) + \varphi'(g(a-))}{2} - \frac{\delta(\varphi \circ g)}{\delta g}(a) \right].$$

Here $J_g \subset [0,1]$ denotes the set of jump locations of g and

$$\frac{\delta(\varphi \circ g)}{\delta g}(a) := \frac{\varphi(g(a+)) - \varphi(g(a-))}{g(a+) - g(a-)}.$$

By formula (4.11) one can extend ∇ to a closed operator on $D(\mathcal{E})$ such that $\mathcal{E}(u,u) = \|\nabla u\|_T^2$. The Markov uniqueness of $\mathcal{E}$ now implies the converse which is a characterization of $D(\mathcal{E})$ via (4.11). For this we need the following technical lemma.

Lemma 4.19. *The functional*

$$\tilde{\mathcal{E}}(u,u)^{1/2} = \sup_{\zeta \in \Theta} \frac{-\langle u, \text{div}_{\mathbb{Q}^\beta} \zeta \rangle_H}{\|\zeta\|_T} \quad \text{on} \quad D(\tilde{\mathcal{E}}) = \{u \in L^2(\mathcal{G}, \mathbb{Q}^\beta) \,|\, \tilde{\mathcal{E}}(u) < \infty\} \quad (4.12)$$

is a Dirichlet form on $L^2(\mathcal{G}, \mathbb{Q}^\beta)$ extending $\mathcal{E}$, i.e. $D(\mathcal{E}) \subset D(\tilde{\mathcal{E}})$ and $\tilde{\mathcal{E}}(u) = \mathcal{E}(u)$ for $u \in D(\mathcal{E})$.

Proof. First note that it makes no difference to (4.12) if Θ is replaced by the larger set $\tilde{\Theta} = \{\xi \,|\, \xi(g,s) = w(g)\varphi(g(s)), w \in \mathfrak{C}^1(\mathcal{G}), \varphi \in \mathcal{C}^\infty([0,1]), \phi(0) = \phi(1) = 0\}$. Obviously, $\tilde{\mathcal{E}}$ is lower semicontinuous in $L^2(\mathcal{G}, \mathbb{Q}^\beta)$ and therefore closed. Moreover, $\tilde{\mathcal{E}}$ is an extension of $\mathcal{E}$, due to (4.11). It remains to show that $\tilde{\mathcal{E}}$ is Markovian. We use the stronger quasi-invariance of $\mathbb{Q}^\beta$ under certain transformations of $\mathcal{G}$, cf. [66, Theorem 4.3]. Let $h \in \mathcal{G}$ be a C^2-diffeomorphisms of $[0,1]$ and let $\tau_h : \mathcal{G} \to \mathcal{G}$, $\tau_h(g) = h \circ g$, then

$$\frac{d(\tau_{h^{-1}})_* \mathbb{Q}^\beta}{d\mathbb{Q}^\beta}(g) = X_h^\beta(g) Y_h^0(g). \quad (4.13)$$

where

$$X_h : \mathcal{G} \to \mathbb{R}; \quad X_h(g) = \exp(\int_0^1 \log h'(g(s))ds)$$

and

$$Y_h^0(g) := \prod_{a \in J_g} \frac{\sqrt{h'(g(a+)) \cdot h'(g(a-))}}{\frac{\delta(h \circ g)}{\delta g}(a)} \frac{1}{\sqrt{h'(g(0)) \cdot h'(g(1-))}}.$$

Given $\zeta = w(\cdot)\phi(\cdot) \in \Theta$ we may apply formula (4.13) in the case when $h = h_t^\phi$, where $\mathbb{R} \times [0,1] \to [0,1], (t,x) \to h_t(x)$ is the flow of smooth diffeomorphisms of $[0,1]$ induced from the ODE

4.4 Identification of the Limit

$\dot{h}_t(x) = \phi(h_t(x))$ with initial condition $h_0(x) = x$. In particular, h_0 is the identity and $h_t^{-1} = h_{-t}$ for all $t \in \mathbb{R}$ by the flow property. Then, arguing as in Section 5 in [66], we have

$$\int_{\mathcal{G}} \frac{u(h_t(g)) - u(g)}{t} \cdot w(g) \mathbb{Q}^\beta(dg) = \frac{1}{t} \int_{\mathcal{G}} \left[u(g) \, w(h_t^{-1}(g)) X^\beta_{h_t^{-1}} Y^0_{h_t^{-1}} - u(g) w(g) \right] \mathbb{Q}^\beta(dg)$$

$$= \frac{1}{t} \int_{\mathcal{G}} u(g) \left[w(h_{-t}(g)) - w(g) \right] \mathbb{Q}^\beta(dg)$$

$$+ \frac{1}{t} \int_{\mathcal{G}} u(g) \, w(g) \left[X^\beta_{h_{-t}} Y^0_{h_{-t}} - 1 \right] \mathbb{Q}^\beta(dg)$$

$$+ \frac{1}{t} \int_{\mathcal{G}} u(g) \, w(g) \left[w(h_{-t}(g)) - w(g) \right] \left[X^\beta_{h_{-t}} Y^0_{h_{-t}} - 1 \right] \mathbb{Q}^\beta(dg).$$

We recall that by Lemma 5.7 in [66]

$$\lim_{t \to 0} \frac{1}{t} \left[X^\beta_{h_t}(g) Y^0_{h_t}(g) - 1 \right] = \frac{\partial}{\partial t} \left[X^\beta_{h_t} Y^0_{h_t} \right]\bigg|_{t=0} = V^\beta_\varphi(g).$$

Together with the fact that the approximations of the logarithmic derivative of $X^\beta_{h_t} Y^0_{h_t}$ w.r.t. the variable t stay uniformly bounded, more precisely $|\log[X^\beta_{h_t} Y^0_{h_t}]| \leq C\,|t|$ for $|t| \leq 1$ (cf. [66, Section 5]), this yields by the dominated convergence theorem

$$\lim_{t \to 0} \int_{\mathcal{G}} \frac{u(h_t(g)) - u(g)}{t} \cdot w(g) \, \mathbb{Q}^\beta(dg) = - \int_{\mathcal{G}} u \, \mathrm{div}_{\mathbb{Q}^\beta} \, \zeta(g) \, \mathbb{Q}^\beta(dg).$$

for any $u \in D(\tilde{\mathcal{E}})$. More precisely, $u \in L^2(\mathcal{G}, \mathbb{Q}^\beta)$ belongs to $D(\tilde{\mathcal{E}})$ if and only if

$$D_\phi u(g) := \mathrm{w\text{-}lim}_{t \to 0} \frac{u(h_t(g)) - u(g)}{t} \text{ exists in } L^2(\mathcal{G}, \mathbb{Q}^\beta)$$

for all $\phi \in C^\infty([0,1])$ with $\phi(0) = \phi(1) = 0$ and by Riesz's representation theorem

$\exists Du \in T$ s.th. $\langle Du, \zeta \rangle_T = -\langle u, \mathrm{div}_{\mathbb{Q}^\beta} \zeta \rangle_{L^2(\mathcal{G}, \mathbb{Q}^\beta)} = \langle D_\phi u, w \rangle_{L^2(\mathcal{G}, \mathbb{Q}^\beta)} \quad \forall \zeta = w(\cdot) \phi(\cdot) \in \tilde{\Theta}$.

Moreover, $\tilde{\mathcal{E}}(u,u) = \|Du\|_T^2$. Now for $u \in D(\tilde{\mathcal{E}})$ and $\kappa \in C^1(\mathbb{R})$ by Taylor's formula $\kappa(u(h_t(g))) - \kappa(u(g)) = \kappa'(\theta) \cdot (u(g_t) - u(g))$ for some $\theta \to u(g)$ for $t \to 0$ such that in case of $\|\kappa'\|_\infty \leq 1$

$$D_\phi \kappa \circ u(g) = \mathrm{w\text{-}lim}_{t \to 0} \frac{\kappa \circ u(h_t(g)) - \kappa \circ u(g)}{t} = \kappa' \circ u(g) \cdot D_\phi u(g) \text{ in } L^2(\mathcal{G}, \mathbb{Q}^\beta).$$

Choose some sequence $u_k \in \mathfrak{C}^1(G)$ such that $u_k(g) \to u(g)$ for $\mathbb{Q}^\beta$-almost all $g \in \mathcal{G}$. Then $\tilde{\zeta}_k = \kappa' \circ u_k(\cdot) \cdot w(\cdot) \phi(\cdot) \in \Theta$. Since $D_\phi u$ and w both belong to $L^2(\mathcal{G}, \mathbb{Q}^\beta)$, by dominated convergence

$$\langle \kappa' \circ u \cdot D_\phi u, w \rangle_{L^2(\mathcal{G}, \mathbb{Q}^\beta)} = \lim_k \langle \kappa' \circ u_k \cdot D_\phi u, w \rangle_{L^2(\mathcal{G}, \mathbb{Q}^\beta)}$$

$$= \lim_k \langle Du, \tilde{\zeta}_k \rangle_T \leq \tilde{\mathcal{E}}^{1/2}(u,u) \lim_k \|\tilde{\zeta}_k\|_T$$

$$\leq \tilde{\mathcal{E}}^{1/2}(u,u) \, \|\zeta\|_T \, .$$

Hence $\kappa \circ u$ in $D(\tilde{\mathcal{E}})$ and $\tilde{\mathcal{E}}(\kappa \circ u) \leq \tilde{\mathcal{E}}(u)$. Applied to a uniformly bounded family $\kappa_\epsilon : \mathbb{R} \to \mathbb{R}$ which converges pointwise to $\kappa(s) = \min(\max(s,0),1)$ the lower semicontinuity of $\tilde{\mathcal{E}}$ yields the claim. □

Corollary 4.20 (Meyers-Serrin property). *Assume Markov-uniqueness holds for $\mathcal{E}$, then*

$$(\mathcal{E}(u,u))^{1/2} = \sup_{\zeta \in \Theta} \frac{-\langle u, \mathrm{div}_{\mathbb{Q}^\beta} \zeta \rangle_H}{\|\zeta\|_T}. \tag{4.14}$$

Proof. The assumption means that $\mathcal{E}$ has no proper extension in the class of Dirichlet forms on $L^2(\mathcal{G}, \mathbb{Q}^\beta)$. By the previous lemma we obtain $\tilde{\mathcal{E}} = \mathcal{E}$ which is the claim. □

The convergence of (4.10) to (4.11) is established by the following lemma whose proof is given below.

Lemma 4.21. *For $\zeta \in \Theta$ there exists a sequence of vector fields $\zeta_N : \overline{\Sigma}_N \to \mathbb{R}^N$ such that $\mathrm{div}_{q_N} \zeta_N \in H^N$ converges strongly to $\mathrm{div}_{\mathbb{Q}^\beta} \zeta$ in H and such that $\|\zeta_N\|_{T^N} \to \|\zeta\|_T$ for $N \to \infty$.*

Proposition 4.22 (Mosco I). *Let $\mathcal{E}$ be Markov-unique and let $u_N \in \mathcal{D}(\mathcal{E}^N)$ converge weakly to $u \in H$, then*

$$\mathcal{E}(u,u) \leq \liminf_{N \to \infty} N \cdot \mathcal{E}^N(u_N, u_N).$$

Proof. Let $u \in H$ and $u_N \in H^N$ converge weakly to u. Let $\zeta \in \Theta$ and ζ^N be as in Lemma 4.21, then

$$-\langle u, \mathrm{div}_{\mathbb{Q}^\beta} \zeta \rangle_H = -\lim \langle u_N, \mathrm{div}_{q_N} \zeta_N \rangle_{H^N} = \lim N \cdot \langle \nabla u_N, \zeta_N \rangle_{T^N}$$

$$\leq \liminf N \cdot \|\nabla u_N\|_{T^N} \cdot \|\zeta_N\|_{T_N} = \liminf \left(N \cdot \mathcal{E}^N(u_N, u_N)\right)^{1/2} \cdot \|\zeta\|_T,$$

such that, using (4.14),

$$(\mathcal{E}(u,u))^{1/2} = \sup_{\zeta \in \Theta} \frac{-\langle u, \mathrm{div}_{\mathbb{Q}^\beta} \zeta \rangle_H}{\|\zeta\|_T} \leq \liminf \left(N \cdot \mathcal{E}^N(u_N, u_N)\right)^{1/2}. \tag{4.15}$$

□

Proof of Lemma 4.21. By linearity it suffices to consider the case $\zeta(g,t) = w(g) \cdot \varphi(g(t))$ with $w(g) = \prod_{i=1}^n l_{f_i}^{k_i}(g)$. Choose

$$(\zeta_N(x^1, \cdots, x^N))^i := w_N(x^1, \ldots, x^N) \cdot \varphi(x^i)$$

with $w_N := \prod_{i=1}^n (\Phi^N(l_{f_i}))^{k_i}$. Then, by Remark 3.8

$$\mathrm{div}_{q_N} \zeta_N = w_N \cdot V_{N,\varphi}^\beta + \langle \nabla w_N, \vec{\varphi} \rangle_{\mathbb{R}^N},$$

with

$$\vec{\varphi}(x^1, \ldots, x^N) := (\varphi(x^1), \ldots, \varphi(x^N))$$

and

$$V_{N,\varphi}^\beta(x^1, \ldots, x^N) := \left(\frac{\beta}{N+1} - 1\right) \sum_{i=0}^N \frac{\varphi(x^{i+1}) - \varphi(x^i)}{x^{i+1} - x^i} + \sum_{i=1}^N \varphi'(x^i).$$

We recall that for all bounded measurable $u : [0,1]^N \to \mathbb{R}$

$$\int_{\Sigma_N} u(x^1, \ldots, x^N) \, q_N(dx) = \int_{\mathcal{G}} u(g(t_1), \ldots, g(t_N)) \, \mathbb{Q}^\beta(dg),$$

with $t_i = i/(N+1)$, $i = 0, \ldots, N+1$. Using this we get immediately

$$\|\zeta_N\|_{T^N}^2 = \frac{1}{N} \int_{\Sigma_N} \sum_{i=1}^N w_N^2(x) \, \varphi(x^i)^2 \, q_N(dx) = \int_{\mathcal{G}} w_N^2(g(t_1), \ldots, g(t_N)) \frac{1}{N} \sum_{i=1}^N \varphi(g(t_i))^2 \, \mathbb{Q}^\beta(dg)$$

$$\to \int_{\mathcal{G}} w^2(g) \int_0^1 \varphi(g(s))^2 \, ds \, \mathbb{Q}^\beta(dg) = \|\zeta\|_T^2.$$

To prove strong convergence of $\mathrm{div}_{q_N} \zeta_N$ to $\mathrm{div}_{\mathbb{Q}^\beta} \zeta$, by definition we have to show that there exists a sequence $(d_N \zeta)_N \subset H$ tending to $\mathrm{div}_{\mathbb{Q}^\beta} \zeta$ in H such that

$$\lim_N \limsup_M \left\| \Phi^M(d_N \zeta) - \mathrm{div}_{q_M} \zeta_M \right\|_{H^M}^2 = 0.$$

The choice

$$d_N \zeta(g) := \mathrm{div}_{q_N} \zeta_N(g(t_1), \ldots, g(t_N))$$

makes this convergence trivial, once we have proven that in fact $(d_N \zeta)_N$ converges to $\mathrm{div}_{\mathbb{Q}^\beta} \zeta$ in H. This is carried out in the following two lemmas.

Lemma 4.23. *For $\mathbb{Q}^\beta$-a.s. g we have*

$$V_{N,\varphi}^\beta(g(t_1), \ldots, g(t_N)) \to V_\varphi^\beta(g), \quad \text{as } N \to \infty,$$

and we have also convergence in $L^p(\mathcal{G}, \mathbb{Q}^\beta)$, $p > 1$.

Proof. We rewrite $V_{N,\varphi}^\beta(g(t_1), \ldots, g(t_N))$ as

$$V_{N,\varphi}^\beta(g(t_1), \ldots, g(t_N)) = \beta \sum_{i=0}^N \frac{\varphi(g(t_{i+1})) - \varphi(g(t_i))}{g(t_{i+1}) - g(t_i)} (t_{i+1} - t_i)$$

$$- \frac{\varphi(g(t_1)) - \varphi(g(t_0))}{g(t_1) - g(t_0)} + \sum_{i=1}^{N-1} \left(\varphi'(g(t_i)) - \frac{\varphi(g(t_{i+1})) - \varphi(g(t_i))}{g(t_{i+1}) - g(t_i)} \right)$$

$$+ \varphi'(g(t_N)) - \frac{\varphi(g(t_{N+1})) - \varphi(g(t_N))}{g(t_{N+1}) - g(t_N)}.$$
(4.16)

Note that all terms are uniformly bounded in g with a bound depending on the supremum norm of φ' and φ'', respectively. Since the same holds for $V_\varphi^\beta(g)$ (cf. Section 5 in [66]), it is sufficient to show convergence $\mathbb{Q}^\beta$-a.s. By the support properties of $\mathbb{Q}^\beta$ g is continuous at $t_{N+1} = 1$, so that the last line in (4.16) tends to zero. Using Taylor's formula we obtain that the first term in (4.16) is equal to

$$\beta \sum_{i=0}^{N} \varphi'(g(t_i))(t_{i+1} - t_i) + \frac{1}{2} \sum_{i=0}^{N} \varphi''(\gamma_i) \left(g(t_{i+1}) - g(t_i)\right)(t_{i+1} - t_i),$$

for some $\gamma_i \in [g(t_i), g(t_{i+1})]$. Obviously, the first term tends to $\beta \int_0^1 \varphi'(g(s)) \, ds$ and the second one to zero as $N \to \infty$. Thus, it remains to show that the second line in (4.16) converges to

$$\sum_{a \in J_g} \left[\frac{\varphi'(g(a+)) + \varphi'(g(a-))}{2} - \frac{\delta(\varphi \circ g)}{\delta g}(a) \right] - \frac{\varphi'(0) + \varphi'(1)}{2}. \quad (4.17)$$

Note that by the right-continuity of g the first term in the second line in (4.16) tends to $-\varphi'(0)$. Let now $a_2, \ldots, a_{l-1}$ denote the $l-2$ largest jumps of g on $]0,1[$. For N very large (compared with l) we may assume that $a_2, \ldots, a_{l-2} \in]\frac{2}{N+1}, 1 - \frac{2}{N+1}[$. Put $a_1 := \frac{1}{N+1}$, $a_l := 1 - \frac{1}{N+1}$. For $j = 1, \ldots, l$ let k_j denote the index $i \in \{1, \ldots, N\}$, for which $a_j \in [t_i, t_{i+1}[$. In particular, $k_1 = 1$ and $k_l = N$. Then

$$\sum_{i \in \{k_2, \ldots, k_{l-1}\}} \varphi'(g(t_i)) - \frac{\varphi(g(t_{i+1})) - \varphi(g(t_i))}{g(t_{i+1}) - g(t_i)} \xrightarrow[N \to \infty]{} \sum_{j=2}^{l-1} \varphi'(g(a_j-)) - \frac{\delta(\varphi \circ g)}{\delta g}(a_j)$$

$$\xrightarrow[l \to \infty]{} \sum_{a \in J_g} \varphi'(g(a-)) - \frac{\delta(\varphi \circ g)}{\delta g}(a). \quad (4.18)$$

Provided l and N are chosen so large that

$$|g(t_{i+1}) - g(t_i)| \leq \frac{C}{l}$$

for all $i \in \{0, \ldots, N\} \setminus \{k_1, \ldots, k_l\}$, where $C = \sup_s |\varphi'''(s)|/6$, again by Taylor's formula we get for every $j \in \{1, \ldots, l-1\}$

$$\sum_{i=k_j+1}^{k_{j+1}-1} \varphi'(g(t_i)) - \frac{\varphi(g(t_{i+1})) - \varphi(g(t_i))}{g(t_{i+1}) - g(t_i)}$$

$$= -\sum_{i=k_j+1}^{k_{j+1}-1} \frac{1}{2} \varphi''(g(t_i)) \left(g(t_{i+1}) - g(t_i)\right) + \frac{1}{6} \varphi'''(\gamma_i) \left(g(t_{i+1}) - g(t_i)\right)^2$$

$$\xrightarrow[N \to \infty]{} -\frac{1}{2} \int_{a_j+}^{a_{j+1}-} \varphi''(g(s)) \, dg(s) + O(l^{-2}) = -\frac{1}{2} \int_{g(a_j+)}^{g(a_{j+1}-)} \varphi''(s) \, ds + O(l^{-2}).$$

4.4 IDENTIFICATION OF THE LIMIT

Summation over j leads to

$$\sum_{j=1}^{l-1}\sum_{i=k_j+1}^{k_{j+1}-1}\varphi'(g(t_i)) - \frac{\varphi(g(t_{i+1})) - \varphi(g(t_i))}{g(t_{i+1}) - g(t_i)}$$

$$\xrightarrow[N\to\infty]{} -\frac{1}{2}\sum_{j=1}^{l-1}\int_{g(a_j+)}^{g(a_{j+1}-)}\varphi''(s)\,ds + O(l^{-1}) = -\frac{1}{2}\int_0^1 \varphi''(s)\,ds + \frac{1}{2}\sum_{j=2}^{l-1}\int_{g(a_j-)}^{g(a_j+)}\varphi''(s)\,ds + O(l^{-1})$$

$$\xrightarrow[l\to\infty]{} -\frac{1}{2}(\varphi'(1) - \varphi'(0)) + \frac{1}{2}\sum_{a\in J_g}\varphi'(g(a+)) - \varphi'(g(a-)).$$

Combining this with (4.18) yields that the second line of (4.16) converges in fact to (4.17), which completes the proof. □

Since $w_N(g(t_1),\ldots,g(t_N))$ converges to w in $L^p(\mathcal{G}, \mathbb{Q}^\beta)$, $p > 0$ (cf. proof of Lemma 4.13 above), the last lemma ensures that the first term of $d_N\zeta$ converges to the first term of $\mathrm{div}_{\mathbb{Q}^\beta}\zeta$ in H, while the following lemma deals with the second term.

Lemma 4.24. *For $\mathbb{Q}^\beta$-a.s. g we have*

$$\langle \nabla w_N(g(t_1),\ldots,g(t_N)), \vec{\varphi}(g(t_1),\ldots,g(t_N))\rangle_{\mathbb{R}^N} \to \langle \nabla w_{|g}, \varphi(g(.))\rangle_{L^2(0,1)}, \quad \text{as } N \to \infty,$$

and we have also convergence in H.

Proof. As in the proof of the last lemma it is enough to prove convergence $\mathbb{Q}^\beta$-a.s. Note that

$$\langle \nabla w_N(\vec{g}), \vec{\varphi}(\vec{g})\rangle_{\mathbb{R}^N} = (N+1)\langle \iota^N(\nabla w_N(\vec{g})), \iota^N(\vec{\varphi}(\vec{g}))\rangle_{L^2(0,1)},$$

writing $\vec{g} := (g(t_1),\ldots,g(t_N))$ and using the extension of ι^N on $\mathbb{R}^N$. By triangle and Cauchy-Schwarz inequality we obtain

$$|\langle (N+1)\iota^N(\nabla w_N(\vec{g})), \iota^N(\vec{\varphi}(\vec{g}))\rangle_{L^2(0,1)} - \langle \nabla w_{|g}, \varphi(g(.))\rangle_{L^2(0,1)}|$$
$$\leq |\langle (N+1)\iota^N(\nabla w_N(\vec{g})) - \nabla w_{|g}, \iota^N(\vec{\varphi}(\vec{g}))\rangle_{L^2(0,1)}| + |\langle \nabla w_{|g}, \iota^N(\vec{\varphi}(\vec{g})) - \varphi(g(.))\rangle_{L^2(0,1)}|$$
$$\leq \left\|(N+1)\iota^N(\nabla w_N(\vec{g})) - \nabla w_{|g}\right\|_{L^2(0,1)} \left\|\iota^N(\vec{\varphi}(\vec{g}))\right\|_{L^2(0,1)}$$
$$+ \left\|\nabla w_{|g}\right\|_{L^2(0,1)} \left\|\iota^N(\vec{\varphi}(\vec{g})) - \varphi(g(.))\right\|_{L^2(0,1)},$$

which tends to zero by Lemma 4.16 and by the definition of ι^N. □

4.4.5 Proof of Proposition 4.4

Lemma 4.25. *For $u \in C(\mathcal{G})$ let $u_N \in H^N$ be defined by $u_N(x) := u(\iota x)$, then $u_N \to u$ strongly. Moreover, for any sequence $f_N \in H^N$ with $f_N \to f \in H$ strongly, $u_N \cdot f_N \to u \cdot f$ strongly.*

Proof. Let $\tilde{u}_N \in H$ be defined by $\tilde{u}_N(g) := u(g^N)$, where $g^N := \sum_{i=1}^{N} g(t_i) \mathbb{1}_{[t_i, t_{i+1})}$, $t_i := i/(N+1)$, then $\tilde{u}_N \to u$ in H strongly. Moreover,

$$\lim_N \lim_M \left\| \Phi^M \tilde{u}_N - u_M \right\|_{H^M} = \lim_N \lim_M \left\| \Phi^M \tilde{u}_N - \tilde{u}_M \right\|_H = \lim_N \|\tilde{u}_N - u\|_H = 0,$$

where as above we have identified Φ^M with the corresponding projection operator in $L^2(\mathcal{G}, \mathbb{Q}^\beta)$. For the proof of the second statement, let $H \ni \tilde{f}_N \to f$ in H such that

$$\lim_N \limsup_M \left\| \Phi^M \tilde{f}_N - f_M \right\|_{H^M} = 0.$$

From the uniform boundedness of $\tilde{u}_N$ it follows that also $\tilde{u}_N \cdot \tilde{f}_N \to u \cdot f$ in H. In order to show $H^M \ni u_M \cdot f_M \to u \cdot f$ write

$$\left\| \Phi^M(\tilde{u}_N \cdot \tilde{f}_N) - u_M \cdot f_M \right\|_{H^M}$$
$$\leq \left\| \Phi^M(\tilde{u}_N \cdot \tilde{f}_N) - u_M \cdot \Phi^M(\tilde{f}_M) \right\|_{H^M} + \left\| u_M \cdot f_M - u_M \cdot \Phi^M(\tilde{f}_M) \right\|_{H^M}.$$

Identifying the map Φ^M with the associated conditional expectation operator, considered as an orthogonal projection in H, the claim follows from

$$\left\| \Phi^M(\tilde{u}_N \cdot \tilde{f}_N) - u_M \cdot \Phi^M(\tilde{f}_M) \right\|_{H^M} = \left\| \Phi^M(\tilde{u}_N \cdot \tilde{f}_N) - \tilde{u}_M \cdot \Phi^M(\tilde{f}_M) \right\|_H$$
$$= \left\| \Phi^M(\tilde{u}_N \cdot \tilde{f}_N) - \Phi^M(\tilde{u}_M \cdot \tilde{f}_M) \right\|_H$$
$$\leq \left\| \tilde{u}_N \cdot \tilde{f}_N - \tilde{u}_M \cdot \tilde{f}_M \right\|_H$$

and

$$\left\| u_M \cdot f_M - u_M \cdot \Phi^M(\tilde{f}_M) \right\|_{H^M} \leq \|u\|_\infty \left\| f_M - \Phi^M(\tilde{f}_N) \right\|_{H^M}$$
$$+ \|u\|_\infty \left\| \Phi^M(\tilde{f}_N) - \Phi^M(\tilde{f}_M) \right\|_{H^M}$$
$$= \|u\|_\infty \left\| f_M - \Phi^M(\tilde{f}_N) \right\|_{H^M}$$
$$+ \|u\|_\infty \left\| \Phi^M(\tilde{f}_N) - \Phi^M(\tilde{f}_M) \right\|_H$$
$$\leq \|u\|_\infty \left\| f_M - \Phi^M(\tilde{f}_N) \right\|_{H^M}$$
$$+ \|u\|_\infty \left\| \tilde{f}_N - \tilde{f}_M \right\|_H$$

such that in fact $\lim_N \limsup_M \left\| \Phi^M(\tilde{u}_N \cdot \tilde{f}_N) - u_M \cdot f_M \right\|_{H^M} = 0$. □

Proof of Proposition 4.4. It suffices to prove the claim for functions $f \in C(\mathcal{G}^l)$ of the form $f(g_1, \ldots, g_l) = f_1(g_1) \cdot f_2(g_2) \cdots f_l(g_l)$ with $f_i \in C(\mathcal{G})$. Let $P_t^N : H^N \to H^N$ be the semigroup on H^N induced by $(X_{N \cdot t}^N)_t$ via $\mathbb{E}_{g \cdot q_N}[f(X_{N \cdot t}^N)] = \langle P_t^N f, g \rangle_{H^N}$. From Theorem 4.8 and the abstract results in [50] it follows that P_t^N converges to P_t strongly, i.e. for any sequence

$u^N \in H^N$ converging to some $u \in H$ strongly, the sequence $P_t^N u^N$ also strongly converges to $P_t u$. Let $f_i^N := f_i \circ \iota^N$, then inductive application of Lemma 4.25 yields

$$P_{t_l - t_{l-1}}^N (f_l^N \cdot P_{t_{l-1} - t_{l-2}}^N (f_{l-1}^N \cdot P_{t_{l-2} - t_{l-3}}^N \cdots f_2^N \cdot P_{t_1}^N f_1^N) \cdots)$$

$$\stackrel{N \to \infty}{\longrightarrow} P_{t_l - t_{l-1}} (f_l \cdot P_{t_{l-1} - t_{l-2}} (f_{l-1} \cdot P_{t_{l-2} - t_{l-3}} \cdots f_2 \cdot P_{t_1} f_1) \cdots) \text{ strongly},$$

which in particular implies the convergence of inner products. Hence, using the Markov property of g^N and g we may conclude that

$$\lim_N \mathbb{E}(f_1(g_{t_1}^N) \cdots f_l(g_{t_l}^N)) = \lim_N \mathbb{E}(f_1^N(X_{N \cdot t_1}^N) \cdots f_l^N(X_{N \cdot t_l}^N))$$
$$= \lim_N \langle 1, P_{t_l - t_{l-1}}^N (f_l^N \cdot P_{t_{l-1} - t_{l-2}}^N (f_{l-1}^N \cdot P_{t_{l-2} - t_{l-3}}^N \cdots f_2^N \cdot P_{t_1}^N f_1^N) \cdots) \rangle_{H^N}$$
$$= \langle 1, P_{t_l - t_{l-1}} (f_l \cdot P_{t_{l-1} - t_{l-2}} (f_{l-1} \cdot P_{t_{l-2} - t_{l-3}} \cdots f_2 \cdot P_{t_1} f_1) \cdots) \rangle_H$$
$$= \mathbb{E}(f_1(g_{t_1}) \cdots f_l(g_{t_l})). \tag{4.19}$$

□

4.5 A non-convex $(1+1)$-dimensional $\nabla \phi$-interface model

We conclude with a remark on a link to stochastic interface models, cf. [40]. Consider an interface on the one-dimensional lattice $\Gamma_N := \{1, \ldots, N\}$, whose location at time t is represented by the height variables $\phi_t = \{\phi_t(x), x \in \Gamma_N\} \in \sqrt{N} \cdot \overline{\Sigma}_N$ with dynamics determined by the generator $\tilde{L}^N$ defined below and with the boundary conditions $\phi_t(0) = 0$ and $\phi(N+1) = \sqrt{N}$ at $\partial \Gamma_N := \{0, N+1\}$.

$$\tilde{L}^N f(\phi) := \left(\frac{\beta}{N+1} - 1\right) \sum_{x \in \Gamma_N} \left(\frac{1}{\phi(x) - \phi(x-1)} - \frac{1}{\phi(x+1) - \phi(x)}\right) \frac{\partial}{\partial \phi(x)} f(\phi) + \Delta f(\phi)$$

for $\phi \in \sqrt{N} \cdot \Sigma_N$ and with $\phi(0) := 0$ and $\phi(N+1) := \sqrt{N}$. $\tilde{L}^N$ corresponds to L^N as an operator on $C^2(\sqrt{N} \cdot \overline{\Sigma}_N)$ with Neumann boundary conditions. Note that this system involves a non-convex interaction potential function V on $(0, \infty)$ given by $V(r) = (1 - \frac{\beta}{N+1}) \log(r)$ and the Hamiltonian

$$H_N(\phi) := \sum_{x=0}^{N} V(\phi(x+1) - \phi(x)), \qquad \phi(0) := 0, \phi(N+1) := \sqrt{N}.$$

Then, the natural stationary distribution of the interface is the Gibbs measure μ_N conditioned on $\sqrt{N} \cdot \overline{\Sigma}_N$:

$$\mu_N(d\phi) := \frac{1}{Z_N} \exp(-H_N(\phi)) \mathbb{1}_{\{(\phi(1), \ldots, \phi(N)) \in \sqrt{N} \cdot \overline{\Sigma}_N\}} \prod_{x \in \Gamma_N} d\phi(x),$$

where Z_N is a normalization constant. Note that μ_N is the corresponding measure of q_N on the state space $\sqrt{N} \cdot \overline{\Sigma}_N$. Suppose now that $(\phi_t)_{t \geq 0}$ is the stationary process generated by $\tilde{L}^N$. Then the space-time scaled process

$$\tilde{\Phi}_t^N(x) := \frac{1}{\sqrt{N}} \phi_{N^2 t}(x), \qquad x = 0, \ldots, N+1,$$

taking values in $\overline{\Sigma}_N$, is associated with the Dirichlet form $N \cdot \mathcal{E}^N$. Introducing the $\mathcal{G}$-valued fluctuation field

$$\Phi_t^N(\vartheta) := \iota^N(\tilde{\Phi}_t^N)(\vartheta) = \sum_{x \in \Gamma_N} \tilde{\Phi}_t^N(x) \, \mathbb{1}_{[x/(N+1),(x+1)/(N+1))}(\vartheta), \qquad \vartheta \in [0,1),$$

by our main result we have weak convergence for the law of the equilibrium fluctuation field Φ^N to the law of the nonlinear diffusion process $\kappa(\mu.)$ on $\mathcal{G}$, which is the $\mathcal{G}$-parametrization of the Wasserstein diffusion.

Appendix A

Conditional Expectation of the Dirichlet Process

This section is devoted to the proof of Lemma 4.14. First we recall some basic facts about the link between Gamma and Dirichlet processes. For $\alpha > 0$ we denote by $G(\alpha)$ the dx-absolutely continuous probability measure on $\mathbb{R}_+$ with density $\frac{1}{\Gamma(\alpha)}x^{\alpha-1}e^{-x}$.

Definition A.1. *A real valued Markov process $(\gamma(t))_{t\geq 0}$ starting in zero is called standard Gamma process if its increments are independent and distributed according to $\gamma(t)-\gamma(s) \sim G(t-s)$ for $0 \leq s < t$.*

As every Lévy process the Gamma process admits a càdlàg modification, i.e. for almost all realizations of γ the function $t \mapsto \gamma(t)$ is càdlàg and nondecreasing. Fix some $T > 0$ and $\beta > 0$. Then, the process $(\frac{\gamma(\beta \cdot t)}{\gamma(\beta \cdot T)})_{t\in[0,T]}$ is called Dirichlet process on $[0,T]$ with parameter β. Its law will be denoted by $\mathbb{Q}^{\beta,T}$ and is obviously concentrated on the set $\mathcal{G}_T := \{g : [0,T) \to [0,1] \,|\, g \text{ càdlàg nondecreasing}\}$. The finite dimensional distributions are given by

$$\mathbb{Q}^{\beta,T}(g(t_1) \in dx^1, \ldots, g(t_N) \in dx^N)$$
$$= \frac{\Gamma(\beta T)}{\prod_{i=0}^{N}\Gamma(\beta(t_{i+1}-t_i))} \prod_{i=0}^{N}(x^{i+1}-x^i)^{\beta(t_{i+1}-t_i)-1} dx^1 \ldots dx^N,$$

for every $N \in \mathbb{N}$ and every partition $0 = t_0 < t_1 < t_2 < \ldots < t_N < t_{N+1} = T$. The precise meaning is that for all bounded measurable $u : [0,1]^N \to \mathbb{R}$,

$$\int_{\mathcal{G}_T} u(g(t_1), \ldots, g(t_N))\, d\mathbb{Q}^{\beta,T}$$
$$= \frac{\Gamma(\beta T)}{\prod_{i=0}^{N}\Gamma(\beta(t_{i+1}-t_i))} \int_{\Sigma_N} u(x^1, \ldots, x^N) \prod_{i=0}^{N}(x^{i+1}-x^i)^{\beta(t_{i+1}-t_i)-1} dx^1 \ldots dx^N,$$

where Σ_N is defined as before. Moreover, it is well known that the Gamma process $(\gamma_{\beta \cdot t})_{t\in[0,T]}$ and its total charge $\gamma(\beta \cdot T)$ are independent (cf. e.g. [32]). In particular, alternatively the law $\mathbb{Q}^{\beta,T}$ of the Dirichlet process can be obtained by conditioning the law of the Gamma process $(\gamma_{\beta \cdot t})_{t\in[0,T]}$

on the event $\gamma(\beta \cdot T) = 1$. In other words, the Dirichlet process on $[0,T]$ can interpreted as some kind of Gamma bridge between zero and T. Finally, we recall that the Dirichlet process is a pure jump process. More precisely, the typical paths are strictly increasing but increase only by jumps and the jump locations are dense in $[0,T]$ (see Section 3 in [66] for more details).

Lemma A.2. *For arbitrary $T > 0$*

$$\mathbb{E}_{\mathbb{Q}^{\beta,T}}[g] = \frac{\mathrm{Id}_{|[0,T]}}{T}.$$

Proof. We need to show that for every partition $0 = t_0 < t_1 < t_2 < \ldots < t_N < t_{N+1} = T$, $\mathbb{E}_{\mathbb{Q}^{\beta,T}}[(g(t_1),\ldots,g(t_N))] = (t_1,\ldots,t_N)/T$, i.e. $\mathbb{E}_{\mathbb{Q}^{\beta,T}}[g(t_i)] = t_i/T$ for every $i \in \{1,\ldots,N\}$. Note that by the formula for the the finite dimensional distributions we have that

$$\mathbb{E}_{\mathbb{Q}^{\beta,T}}[g(t_i)] = \frac{\Gamma(\beta T)}{\prod_{j=0}^{N} \Gamma(\beta(t_{j+1} - t_j))} \int_{\Sigma^N} x^i \prod_{j=0}^{N}(x^{j+1} - x^j)^{\beta(t_{j+1}-t_j)-1} dx^1 \ldots dx^N$$

$$= \frac{\Gamma(\beta T)}{\prod_{j=0}^{N} \Gamma(\beta(t_{j+1} - t_j))} \int_0^1 x^i A(x^i) B(x^i) dx^i$$

with

$$A(x^i) := \int_0^{x^i} \int_0^{x^{i-1}} \cdots \int_0^{x^2} \prod_{j=0}^{i-1}(x^{j+1} - x^j)^{\beta(t_{j+1}-t_j)-1} dx^1 \cdots dx^{i-1},$$

$$B(x^i) := \int_{x^i}^1 \int_{x^{i+1}}^1 \cdots \int_{x^{N-1}}^1 \prod_{j=i}^{N}(x^{j+1} - x^j)^{\beta(t_{j+1}-t_j)-1} dx^N \cdots dx^{i+1}.$$

We will use the well known fact

$$\int_a^b (z-a)^u (b-z)^v\, dz = \frac{\Gamma(1+u)\Gamma(1+v)}{\Gamma(2+u+v)} (b-a)^{1+u+v},$$

for arbitrary $0 \leq a < b$ and $u,v > -1$ to compute $A(x^i)$:

$$A(x^i)$$
$$= \int_0^{x^i} \cdots \int_0^{x^3} \prod_{j=2}^{i-1}(x^{j+1} - x^j)^{\beta(t_{j+1}-t_j)-1} \int_0^{x^2} (x^1)^{\beta t_1 - 1}(x^2 - x^1)^{\beta(t_2-t_1)-1} dx^1\, dx^2 \cdots dx^{i-1}$$

$$= \Gamma(\beta t_1)\, \Gamma(\beta(t_2 - t_1)) \int_0^{x^i} \cdots \int_0^{x^4} \prod_{j=2}^{i-1}(x^{j+1} - x^j)^{\beta(t_{j+1}-t_j)-1}$$
$$\times \frac{1}{\Gamma(\beta t_2)} \int_0^{x^3}(x^2)^{\beta t_2 - 1}(x^3 - x^2)^{\beta(t_3-t_2)-1} dx^2\, dx^3 \cdots dx^{i-1}$$

$$= \Gamma(\beta t_1)\, \Gamma(\beta(t_2 - t_1)) \int_0^{x^i} \cdots \int_0^{x^4} \prod_{j=2}^{i-1}(x^{j+1} - x^j)^{\beta(t_{j+1}-t_j)-1} \frac{\Gamma(\beta(t_3 - t_2))}{\Gamma(\beta t_3)}(x^3)^{\beta t_3 - 1} dx^3 \cdots dx^{i-1}.$$

Iterating this procedure yields

$$A(x^i) = \frac{\prod_{j=0}^{i-1} \Gamma(\beta(t_{j+1} - t_j))}{\Gamma(\beta t_i)} (x^i)^{\beta t_i - 1}.$$

Analogously, we obtain

$$B(x^i) = \frac{\prod_{j=i}^{N} \Gamma(\beta(t_{j+1} - t_j))}{\Gamma(\beta(T - t_i))} (1 - x^i)^{\beta(T - t_i) - 1}.$$

Thus,

$$\mathbb{E}_{\mathbb{Q}^{\beta,T}}[g(t_i)] = \frac{\Gamma(\beta T)}{\Gamma(\beta t_i)\Gamma(\beta(T - t_i))} \int_0^1 (x^i)^{\beta t_i}(1 - x^i)^{\beta(T - t_i) - 1} dx^i$$
$$= \frac{\Gamma(\beta T)}{\Gamma(\beta t_i)\Gamma(\beta(T - t_i))} \cdot \frac{\Gamma(\beta t_i + 1)\Gamma(\beta(T - t_i))}{\Gamma(\beta T + 1)},$$

and using $x\Gamma(x) = \Gamma(x + 1)$, $x > 0$ we finally obtain

$$\mathbb{E}_{\mathbb{Q}^{\beta,T}}[g(t_i)] = \frac{t_i}{T}.$$

$\square$

Lemma A.3. *Let $(\gamma(t))_{t \geq 0}$ denote the Gamma process and let $s > 0$ be arbitrary but fixed. Then, for every $a > 0$ and every measurable $F : \mathcal{G}_s \to \mathbb{R}$*

$$\mathbb{E}\left[F(\gamma_{|[0,s)}) \Big| \gamma(s) = a\right] = \mathbb{E}\left[F(a \cdot \gamma_{|[0,s)}) \Big| \gamma(s) = 1\right].$$

Proof. We set $\tilde{\gamma} := \frac{\gamma}{a}$. In particular note that $\frac{\tilde{\gamma}}{\tilde{\gamma}(s)} = \frac{\gamma}{\gamma(s)}$ is the Dirichlet process over $[0, s]$ (with parameter $\beta = 1$). Hence,

$$\mathbb{E}\left[F(\gamma_{|[0,s)}) \Big| \gamma(s) = a\right] = \mathbb{E}\left[F(a\tilde{\gamma}_{|[0,s)}) \Big| \tilde{\gamma}(s) = 1\right] = \mathbb{E}\left[F\left(a\frac{\tilde{\gamma}_{|[0,s)}}{\tilde{\gamma}(s)}\right) \Big| \tilde{\gamma}(s) = 1\right]$$
$$= \mathbb{E}\left[F\left(a\frac{\tilde{\gamma}_{|[0,s)}}{\tilde{\gamma}(s)}\right)\right] = \mathbb{E}\left[F\left(a\frac{\gamma_{|[0,s)}}{\gamma(s)}\right)\right] = \mathbb{E}\left[F(a \cdot \gamma_{|[0,s)}) \Big| \gamma(s) = 1\right].$$

$\square$

Proof of Lemma 4.14 Setting $t_i = i/(N+1)$, $i = 0, \ldots, N+1$, we have

$$\mathbb{E}_{\mathbb{Q}^\beta}[g|\mathcal{F}_N](X) = \mathbb{E}_{\mathbb{Q}^\beta}\left[\sum_{i=0}^{N} g\mathbb{1}_{[t_i, t_{i+1})} \Big| g(t_1) = x^1, \ldots, g(t_N) = x^N\right]$$
$$= \sum_{i=0}^{N} \mathbb{1}_{[t_i, t_{i+1})} \mathbb{E}_{\mathbb{Q}^\beta}\left[g_{|[t_i, t_{i+1})} \Big| g(t_i) = x^i, g(t_{i+1}) = x^{i+1}\right]$$

Using the stationarity of the Gamma process and Lemma A.3 we obtain for every i

$$\mathbb{E}_{\mathbb{Q}^\beta}\left[g_{|[t_i,t_{i+1})}\Big| g(t_i) = x^i, g(t_{i+1}) = x^{i+1}\right]$$
$$= x^i + \mathbb{E}\left[\gamma(\beta\cdot)_{|[t_i,t_{i+1})}\Big| \gamma(\beta t_i) = 0, \gamma(\beta t_{i+1}) = x^{i+1} - x^i\right]$$
$$= x^i + \mathbb{E}\left[\gamma(\beta(\cdot - t_i))_{|[t_i,t_{i+1})}\Big| \gamma(0) = 0, \gamma(\beta(t_{i+1} - t_i)) = x^{i+1} - x^i\right]$$
$$= x^i + (x^{i+1} - x^i)\,\mathbb{E}\left[\gamma(\beta(\cdot - t_i))_{|[t_i,t_{i+1})}\Big| \gamma(0) = 0, \gamma(\beta(t_{i+1} - t_i)) = 1\right]$$
$$= x^i + (x^{i+1} - x^i)\,\mathbb{E}_{\mathbb{Q}^{\beta,t_{i+1}-t_i}}\left[g(\cdot - t_i)_{|[t_i,t_{i+1})}\right].$$

Finally, by Lemma A.2

$$\mathbb{E}_{\mathbb{Q}^\beta}\left[g|\mathcal{F}_N\right](X) = \sum_{i=0}^{N} \mathbb{1}_{[t_i,t_{i+1})}\left(x^i + (x^{i+1} - x^i)\frac{\cdot - t_i}{t_{i+1} - t_i}\right) = g_X.$$

Bibliography

[1] H. AIRAULT, *Perturbations singulières et solutions stochastiques de problèmes de D. Neumann-Spencer*, J. Math. Pures Appl. (9) **55**, no. 3 (1976), pp. 233–267.

[2] S. ALBEVERIO AND M. RÖCKNER, *Classical Dirichlet forms on topological vector spaces—closability and a Cameron-Martin formula*, J. Funct. Anal. **88**, no. 2 (1990), pp. 395–436.

[3] S. ALBEVERIO AND M. RÖCKNER, *Dirichlet form methods for uniqueness of martingale problems and applications*, in Stochastic analysis (Ithaca, NY, 1993), Amer. Math. Soc., Providence, RI, 1995, pp. 513–528.

[4] A. ALONSO AND F. BRAMBILA-PAZ, L^p-*continuity of conditional expectations*, J. Math. Anal. Appl. **221**, no. 1 (1998), pp. 161–176.

[5] R. F. ANDERSON AND S. OREY, *Small random perturbation of dynamical systems with reflecting boundary*, Nagoya Math. J. **60** (1976), pp. 189–216.

[6] S. ANDRES, *Pathwise differentiability for stochastic differential equations with reflection* (2006). Diploma thesis, Technische Universität Berlin.

[7] S. ANDRES, *Pathwise differentiability for SDEs in a smooth domain with reflection* (2008). Preprint, submitted.

[8] S. ANDRES, *Pathwise differentiability for SDEs in a convex polyhedron with oblique reflection*, Ann. Inst. Henri Poincaré Probab. Stat. **45**, no. 1 (2009), pp. 104–116.

[9] S. ANDRES AND M.-K. VON RENESSE, *Particle approximation of the Wasserstein diffusion* (2008). Preprint, available on arXiv:0712.2387v1 [math.PR].

[10] S. ANDRES AND M.-K. VON RENESSE, *Regularity properties for a system of interacting two-sided bessel processes* (2008). Preprint, in preparation.

[11] R. F. BASS, K. BURDZY, AND Z.-Q. CHEN, *Uniqueness for reflecting Brownian motion in lip domains*, Ann. Inst. H. Poincaré Probab. Statist. **41**, no. 2 (2005), pp. 197–235.

[12] R. F. BASS AND P. HSU, *Some potential theory for reflecting Brownian motion in Hölder and Lipschitz domains*, Ann. Probab. **19**, no. 2 (1991), pp. 486–508.

[13] J. BERTOIN, *Excursions of a* $BES_0(d)$ *and its drift term* $(0 < d < 1)$, Probab. Theory Related Fields **84**, no. 2 (1990), pp. 231–250.

[14] K. BURDZY, *Differentiability of stochastic flow of reflected Brownian motions* (2008). Preprint.

[15] K. BURDZY AND Z.-Q. CHEN, *Coalescence of synchronous couplings*, Probab. Theory Related Fields **123**, no. 4 (2002), pp. 553–578.

[16] K. BURDZY, Z.-Q. CHEN, AND P. JONES, *Synchronous couplings of reflected Brownian motions in smooth domains*, Illinois J. Math. **50**, no. 1-4 (2006), pp. 189–268 (electronic).

[17] K. BURDZY AND J. M. LEE, *Multiplicative functional for reflected Brownian motion via deterministic ODE* (2008). Preprint.

[18] K. L. CHUNG, *Lectures from Markov processes to Brownian motion*, Grundlehren der Mathematischen Wissenschaften 249, Springer-Verlag, New York, 1982.

[19] K. L. CHUNG, *Doubly-Feller process with multiplicative functional*, in Seminar on stochastic processes, 1985 (Gainesville, Fla., 1985), Progr. Probab. Statist. 12, Birkhäuser Boston, Boston, MA, 1986, pp. 63–78.

[20] K. L. CHUNG AND R. J. WILLIAMS, *Introduction to stochastic integration*, Probability and its Applications, Birkhäuser Boston Inc., Boston, MA, second ed., 1990.

[21] K. L. CHUNG AND Z. X. ZHAO, *From Brownian motion to Schrödinger's equation*, Grundlehren der Mathematischen Wissenschaften [Fundamental Principles of Mathematical Sciences] 312, Springer-Verlag, Berlin, 1995.

[22] M. CRANSTON AND Y. LE JAN, *On the noncoalescence of a two point Brownian motion reflecting on a circle*, Ann. Inst. H. Poincaré Probab. Statist. **25**, no. 2 (1989), pp. 99–107.

[23] M. CRANSTON AND Y. LE JAN, *Noncoalescence for the Skorohod equation in a convex domain of $\mathbf{R}^2$*, Probab. Theory Related Fields **87**, no. 2 (1990), pp. 241–252.

[24] D. A. DAWSON, *Measure-valued Markov processes*, in École d'Été de Probabilités de Saint-Flour XXI—1991, Lecture Notes in Math. 1541, Springer, Berlin, 1993, pp. 1–260.

[25] I. V. DENISOV, *A random walk and a Wiener process near a maximum*, Theor. Prob. Appl. **28** (1984), pp. 821–824.

[26] J.-D. DEUSCHEL AND L. ZAMBOTTI, *Bismut-Elworthy's formula and random walk representation for SDEs with reflection*, Stochastic Process. Appl. **115**, no. 6 (2005), pp. 907–925.

[27] M. DÖRING AND W. STANNAT, *The Logarithmic Sobolev Inequality for the Wasserstein Diffusion* (2008). Preprint, to appear in PTRF.

[28] P. DUPUIS AND H. ISHII, *On Lipschitz continuity of the solution mapping to the Skorokhod problem, with applications*, Stochastics Stochastics Rep. **35**, no. 1 (1991), pp. 31–62.

[29] P. DUPUIS AND H. ISHII, *SDEs with oblique reflection on nonsmooth domains*, Ann. Probab. **21**, no. 1 (1993), pp. 554–580.

[30] E. B. DYNKIN, *Markov processes. Vols. I, II*, Die Grundlehren der Mathematischen Wissenschaften 122, Academic Press Inc., Publishers, New York, 1965.

[31] A. EBERLE, *Uniqueness and non-uniqueness of semigroups generated by singular diffusion operators*, Springer-Verlag, Berlin, 1999.

[32] M. ÉMERY AND M. YOR, *A parallel between Brownian bridges and gamma bridges*, Publ. Res. Inst. Math. Sci. **40**, no. 3 (2004), pp. 669–688.

[33] S. N. ETHIER AND T. G. KURTZ, *Markov processes*, Wiley Series in Probability and Mathematical Statistics: Probability and Mathematical Statistics, John Wiley & Sons Inc., New York, 1986. Characterization and convergence.

[34] E. FABES, D. JERISON, AND C. KENIG, *The Wiener test for degenerate elliptic equations*, Ann. Inst. Fourier (Grenoble) **32**, no. 3 (1982), pp. vi, 151–182.

[35] E. B. FABES, C. E. KENIG, AND R. P. SERAPIONI, *The local regularity of solutions of degenerate elliptic equations*, Comm. Partial Differential Equations **7**, no. 1 (1982), pp. 77–116.

Bibliography

[36] T. FATTLER AND M. GROTHAUS, *Strong Feller properties for distorted Brownian motion with reflecting boundary condition and an application to continuous N-particle systems with singular interactions*, J. Funct. Anal. **246**, no. 2 (2007), pp. 217–241.

[37] M. FUKUSHIMA, *A construction of reflecting barrier Brownian motions for bounded domains*, Osaka J. Math. **4** (1967), pp. 183–215.

[38] M. FUKUSHIMA, *On semi-martingale characterizations of functionals of symmetric Markov processes*, Electron. J. Probab. **4** (1999), pp. no. 18, 32 pp. (electronic).

[39] M. FUKUSHIMA, Y. ŌSHIMA, AND M. TAKEDA, *Dirichlet forms and symmetric Markov processes*, Walter de Gruyter & Co., Berlin, 1994.

[40] T. FUNAKI, *Stochastic interface models*, in Lectures on probability theory and statistics, Lecture Notes in Math. 1869, Springer, Berlin, 2005, pp. 103–274.

[41] M. GROTHAUS, Y. G. KONDRATIEV, AND M. RÖCKNER, *N/V-limit for stochastic dynamics in continuous particle systems*, Probab. Theory Related Fields **137**, no. 1-2 (2007), pp. 121–160.

[42] N. IKEDA AND S. WATANABE, *Stochastic differential equations and diffusion processes*, North-Holland Mathematical Library 24, North-Holland Publishing Co., Amsterdam, 1981.

[43] J.-P. IMHOF, *Density factorizations for Brownian motion, meander and the three-dimensional Bessel process, and applications*, J. Appl. Probab. **21**, no. 3 (1984), pp. 500–510.

[44] K. ITÔ AND H. P. MCKEAN, JR., *Diffusion processes and their sample paths*, Springer-Verlag, Berlin, 1974. Second printing, corrected, Die Grundlehren der mathematischen Wissenschaften, Band 125.

[45] I. KARATZAS AND S. E. SHREVE, *Brownian motion and stochastic calculus*, Graduate Texts in Mathematics 113, Springer-Verlag, New York, second ed., 1991.

[46] T. KILPELÄINEN, *Weighted Sobolev spaces and capacity*, Ann. Acad. Sci. Fenn. Ser. A I Math. **19**, no. 1 (1994), pp. 95–113.

[47] C. KIPNIS AND C. LANDIM, *Scaling limits of interacting particle systems*, Springer-Verlag, Berlin, 1999.

[48] A. V. KOLESNIKOV, *Mosco convergence of Dirichlet forms in infinite dimensions with changing reference measures*, J. Funct. Anal. **230**, no. 2 (2006), pp. 382–418.

[49] A. M. KULIK, *Markov uniqueness and Rademacher theorem for smooth measures on infinite-dimensional space under successful filtration condition*, Ukraïn. Mat. Zh. **57**, no. 2 (2005), pp. 170–186.

[50] K. KUWAE AND T. SHIOYA, *Convergence of spectral structures: a functional analytic theory and its applications to spectral geometry*, Comm. Anal. Geom. **11**, no. 4 (2003), pp. 599–673.

[51] P.-L. LIONS AND A.-S. SZNITMAN, *Stochastic differential equations with reflecting boundary conditions*, Comm. Pure Appl. Math. **37**, no. 4 (1984), pp. 511–537.

[52] U. MOSCO, *Composite media and asymptotic Dirichlet forms*, J. Funct. Anal. **123**, no. 2 (1994), pp. 368–421.

[53] K. R. PARTHASARATHY, *Probability measures on metric spaces*, Probability and Mathematical Statistics, No. 3, Academic Press Inc., New York, 1967.

[54] A. PILIPENKO, *Stochastic flows with reflection* (2008). Preprint, available on arXiv:0810.4644.

[55] P. PROTTER, *Stochastic integration and differential equations*, Applications of Mathematics (New York) 21, Springer-Verlag, Berlin, 1990. A new approach.

[56] D. REVUZ AND M. YOR, *Continuous martingales and Brownian motion*, Grundlehren der Mathematischen Wissenschaften 293, Springer-Verlag, Berlin, third ed., 1999.

[57] F. RUSSO, P. VALLOIS, AND J. WOLF, *A generalized class of Lyons-Zheng processes*, Bernoulli **7**, no. 2 (2001), pp. 363–379.

[58] D. W. STROOCK, *Diffusion semigroups corresponding to uniformly elliptic divergence form operators*, in Séminaire de Probabilités, XXII, Lecture Notes in Math. 1321, Springer, Berlin, 1988, pp. 316–347.

[59] K. T. STURM, *Analysis on local Dirichlet spaces. III. The parabolic Harnack inequality*, J. Math. Pures Appl. (9) **75**, no. 3 (1996), pp. 273–297.

[60] H. TANAKA, *Stochastic differential equations with reflecting boundary condition in convex regions*, Hiroshima Math. J. **9**, no. 1 (1979), pp. 163–177.

[61] A. TORCHINSKY, *Real-variable methods in harmonic analysis*, Pure and Applied Mathematics 123, Academic Press Inc., Orlando, FL, 1986.

[62] G. TRUTNAU, *Skorokhod decomposition of reflected diffusions on bounded Lipschitz domains with singular non-reflection part*, Probab. Theory Related Fields **127**, no. 4 (2003), pp. 455–495.

[63] B. O. TURESSON, *Nonlinear potential theory and weighted Sobolev spaces*, Lecture Notes in Mathematics 1736, Springer-Verlag, Berlin, 2000.

[64] S. R. S. VARADHAN AND R. J. WILLIAMS, *Brownian motion in a wedge with oblique reflection*, Comm. Pure Appl. Math. **38**, no. 4 (1985), pp. 405–443.

[65] C. VILLANI, *Topics in optimal transportation*, Graduate Studies in Mathematics 58, American Mathematical Society, Providence, RI, 2003.

[66] M.-K. VON RENESSE AND K.-T. STURM, *Entropic Measure and Wasserstein Diffusion*, Ann. Probab. **37**, no. 3 (2009), pp. 1114–1191.

[67] M.-K. VON RENESSE, M. YOR, AND L. ZAMBOTTI, *Quasi-invariance properties of a class of subordinators* (2007). Preprint, arXiv:0706.3010, to appear in *Stochastic Process. Appl.*

Die VDM Verlagsservicegesellschaft sucht für wissenschaftliche Verlage abgeschlossene und herausragende

Dissertationen, Habilitationen, Diplomarbeiten, Master Theses, Magisterarbeiten usw.

für die kostenlose Publikation als Fachbuch.

Sie verfügen über eine Arbeit, die hohen inhaltlichen und formalen Ansprüchen genügt, und haben Interesse an einer honorarvergüteten Publikation?

Dann senden Sie bitte erste Informationen über sich und Ihre Arbeit per Email an *info@vdm-vsg.de*.

Sie erhalten kurzfristig unser Feedback!

VDM Verlagsservicegesellschaft mbH
Dudweiler Landstr. 99   Telefon +49 681 3720 174
D - 66123 Saarbrücken   Fax   +49 681 3720 1749
www.vdm-vsg.de

Die VDM Verlagsservicegesellschaft mbH vertritt

Printed by Books on Demand GmbH, Norderstedt / Germany